CHARACTERISATION OF HIGH-TEMPERATURE
MATERIALS

Series Editor: M M

# CHEMICAL
# CHARACTERISATION

*Proceedings of the second
seminar in a series of seven
sponsored and organised by the
Materials Science, Materials Engineering and
Continuing Education Committees of
The Institute of Metals; held
in London on 7 September 1988*

EDITED BY I ELLIOTT

THE INSTITUTE OF METALS
1988

Book 444
Published in 1988 by
The Institute of Metals
1 Carlton House Terrace   London SW1Y 5DB

Distributed in North America by
The Institute of Metals   North American Publications Center
Old Post Road   Brookfield   VT 05036 USA

*British Library Cataloguing in Publication Data*

```
Chemical characterisation.
   1. High-temperature materials, Chemical
analysis
   I. Elliott, I.    II. Series
620.1'1217

ISBN 0-901462-50-0
```

*Library of Congress Cataloguing in Publication Data*

*applied for*

Compiled from typematter and illustrations
provided by the authors

Printed and made in England by
Inprint of Luton   Bedfordshire

# PREFACE

The ability to characterise the chemistry of metals has long been one of the corner stones of metallurgy and without it important developmemts such as new alloy systems, microalloying of steels and more recently the ability to design new materials down to the very atomic level would not have been possible. However, despite the close relationship of the two disciplines, the average materials engineer considers chemical characterisation a somewhat distant subject, and with the recent rapid developments in analytical techniques this problem becomes more acute.

The purpose of this volume, the second in the Characterisation of High-temperature Materials series, is to provide those active in the field of high-temperature alloy investigation with a clearer idea of the range of available analytical techniques and particularly their applicability to specific problems. Bulk analysis of steels, nickel-based and Ti alloys are discussed with particular reference to X-ray fluorescence, emission and absorption spectroscopy, as well as the more specific techniques for determination of gases and trace elements. There are contrasted with the use of energy dispersive spectroscopy on the SEM, electron probe microanalysis, surface analysis by SIMS, AES and XPS and some less common techniques employing lasers and accelerators. The authors are drawn from a wide range of industries involved with high-temperature materials and their combined experiences allow an invaluable insight into the practical aspects of chemical characterisation.

Ian Elliott
Head, Research Group,
Inco Alloys Ltd.,
Hereford.

iii

# CONTENTS

All papers are
arranged with
Tables and Figures
following text

# 1 : X-ray fluorescence spectroscopy
# K E WATSON

The author is with Howmet (UK) Inc.

## 1. INTRODUCTION

X-ray spectroscopy is based on the fact that atoms can be made to emit characteristic radiation when stimulated by the appropriate excitation. The various elements each produce their own characteristic line spectra and measurement of these can therefore produce information about the constituents of the sample. The excitation can be produced either by impact of particles such as proton, electron, alpha particles, etc., or by the impact of high energy radiation such as that produced by an X-ray tube. X-ray fluorescence spectroscopy uses the latter method as a means of producing excitation of samples that is both sensitive and versatile enough to be used in practical, routine bulk chemical analysis.

X-ray fluorescence (XRF) spectroscopy was brought in commercially during the 1950's and has been steadily improved and refined since then. The revolution in electronics over the past 20-25 years has resulted in today's highly sophisticated XRF spectrometers that are able to analyse a wide variety of elements over a very wide concentration range and with extremely high precision.

1

## 2. THE X-RAY EMISSION PROCESS

When a sample is bombarded by X-ray photons a number of processes occur, leading to the apparent absorption of the X-ray beam. One of these processes is the transfer of energy between the photon and an electron of one of the constituent atoms. If the energy of the photon is sufficiently great this will lead to the ejection of the electron from the atom. The atom thus becomes unstable and very rapidly electrons transfer from the outer shells to fill the "gap" left by the ejected electron. In transferring to inner shells the electrons lose energy which appears as an emitted X-ray photon. This can either be re-absorbed in the same atom by interaction with other electrons or can escape from the atom to produce an X-ray emission. The overall effect, then, is of absorption of an X-ray beam, followed by emission of a set of secondary X-rays: the phenomenon of X-ray fluorescence.

Fig. 1 shows in a diagrammatic way the ionisation process. In this case an electron from the K-shell has been ejected. Electrons from the L, M and N shells can now make the transition to the partially vacant shell, emitting radiation as they do so. The frequency (and therefore wavelength) of the emission is dependent on the energies of the shell from which the electron starts and the shell in which it finishes. Each possible transition (e.g. L → K, M → K, L → M, etc) therefore produces its own discrete X-ray wavelength, leading to the line spectra observed in X-ray spectroscopy. These inter-shell transitions are further complicated by the fact that all the shells, apart from K, have several sub-shells each with their own energy levels.
Fig. 2 shows part of an energy level diagram and the origin of some of the main X-ray lines.
By convention the X-ray lines are named after the shell in which the parent electronic transition ends. Thus transition to the K shell produces K-lines, etc. A Greek letter is added to differentiate the various lines in each group in a rather non-systematic way leading to the "Siegbahn" notation used by working spectroscopists.
The energy levels of each shell vary from element to element, i.e. the energy of a K-electron in one element is not the same as the energy of a K-electron in

2

another element.  The energy required to eject an
electron from a shell is called the critical excitation
energy.  For example the critical excitation energy for
a K-electron in sodium is 1.072 keV, whilst in
molybdenum it is 20.002 KeV.  The energy emitted on
transition between the shells therefore varies between
the elements.  This means that the wavelength of, say,
the sodium Kα, line is different (at 11.91 Å) from that
of the molybdenum Kα, line (0.71 Å).  In general as we
move up in atomic number the wavelengths become shorter.
This leads to the fact that each element produces its
own unique set of fluorescence wavelengths which can be
measured and used for analysis.

## 3.   INSTRUMENTATION

### 3.1   Spectrometer Layout

The secondary, or fluorescence, radiation produced by
the sample contains all the X-ray wavelengths from the
various electronic transitions.  This is of no use to
the analyst unless it can be separated into its
constituent wavelengths and the intensity of each one
measured.
        The dispersion of the radiation into its
wavelengths is achieved by the use of a crystal placed
at an angle to the X-ray beam.  The radiation is
reflected by the crystal according to the Bragg
equation.

$$\lambda = 2d \sin \theta$$

where   $\lambda$ = wavelength
        d = crystal lattice spacing
        $\theta$ = angle of reflection

From this it can be seen that for any given crystal
the various wavelengths are reflected at different
angles, giving rise to a dispersion of the radiation by
wavelength.  A detector placed at the correct angle to
the crystal will therefore be able to measure any
desired wavelength.
        The layout of a typical spectrometer is shown
schematically in figure 3.  An X-ray tube is used to
produce the primary radiation which is incident on the
flat surface of the sample.  The secondary radiation is
then passed through a collimating device and the

3

resulting beam is directed to the surface of the analysing crystal. This can be made from a variety of materials chosen to give a suitable d-spacing. The most commonly used type is lithium fluoride (LiF) cut on the 200 plane. The reflected radiation is then passed through another collimator and into the detector.

All XRF spectrometers follow this basic pattern although they form two distinct groups within it. First are the so-called sequential spectrometers. These have the crystal and detectors on moveable mounts that can be set to any given angle. They are called sequential since each wavelength must be measured in turn, after positioning on each required angle. The second group are the simultaneous spectrometers. These have a number of crystal/detector combinations each set at a fixed angle, each measuring the intensity of one wavelength, and all operating simultaneously. The sequential instruments are slower than the simultaneous instruments but are more flexible; simultaneous instruments provide greater speed and, as will be seen later, can provide greater precision, but are fixed to measure the pre-set wavelengths. Both instruments find favour in industry: where a routine set of elements must be analysed with maximum speed (such as a steel works) or where high sample throughput is required simultaneous instruments are chosen. Where speed is less important and the elements required for analysis are changing then sequential instruments are preferred. Recently instruments have appeared that combine these two systems thus providing the twin advantages of speed and flexibility.

### 3.1 Detection and Measurement
The detectors used to measure the intensity of the wavelengths also fall into two groups, each covering a specific wavelength range. Scintillation counters are generally used for elements heavier than iron. These make use of the fact that certain types of crystal produce flashes of light (scintillation) when bombarded by X-ray photons. A photomultiplier tube placed next to the crystal will be able to detect these flashes and produce a pulse of electric current that can be amplified and measured. The second type of detector is based on the Geiger-Muller counter in which the X-ray photons pass into a chamber containing a gas. The photons cause ionisation to occur resulting in a pulse

of current between two electrodes in the chamber
(usually a wire and the body of the chamber). Once
again this current can be amplified and measured.

Each detector type has its own response
characteristics. Scintillation counters are better
suited to the shorter wavelengths produced by the
heavier elements. Spectrometers are normally equipped
with both types to ensure the optimum measuring system
is used across the entire wavelength range.

The pulses produced by the X-ray photon in the
detector are processed and counted by electronic
circuitry that includes pulse height filtering to reject
signals from multiple order X-ray lines. The final
output is in the form of a "count rate" expressed as
counts (or kilocounts) per second. This is the raw data
on which the analysis is based.

## 4. APPLICATION

The techniques of X-ray fluorescence spectroscopy can be
applied to a very wide range of elements. The cement
industry routinely analyses the fluorine content of minerals
by XRF, and practically every other element right up to
and including the actinides can be determined. It is
possible to analyse lighter elements such as carbon and
boron but the sensitivity is very low and it is really
only feasible where the concentrations are very high,
such as in borides and carbides.

As well as a wide range of elements XRF also has
the advantage of looking at the innermost electron
shells. The electrons involved are not the ones used in
interatomic bonding and therefore the chemical
state of the atom is not of importance. This means that
an element can be equally well analysed whether it is in
pure form, bound in a molecular compound (e.g. metal
oxides), in solution, or even, in extreme cases, as a
gas.

The physical condition of the sample does not
present a major problem either. Preparation techniques
have been developed to deal with most types of sample.

Direct analysis of solids, such as metals and
alloys is one of the main uses of XRF. The sample is
prepared by cutting, grinding, and the polishing of the
surface. Striations on the surface of the sample
produce an effect called shielding which decreases the
observed fluorescence. This effect is particularly

5

noticeable on lighter elements and is dependent on the angle of the striations to the primary X-ray beam. To help overcome this most spectrometers incorporate a device to spin the sample during analysis to smooth out the effects. It is also important to ensure that the surface finish is consistent between samples and between samples and standards.

Some samples do not lend themselves to polishing or may be in a crushed or powdered form. Examples of this would be oxides and slags. These samples can be treated in two ways, pelletising and fusion.

Pelletising consists of grinding the sample in a mill to a fine particle size, ideally down to 400 mesh, then mixing the powder with a binder such a polyvinyl alcohol. The mixture is then put into a die and pressed into a pellet under hydraulic pressure. The pellet is removed and analysed without further preparation. The fluorescence intensity for a given element can vary with particle size and the amount of pressure used, however with sufficient grinding and a high enough pressure good results can be obtained.

The problems associated with pellets can be overcome by using fusion preparation. In this method the sample is heated in a platinum crucible to around 1000°C with a flux such as lithium tetraborate. The molten flux dissolves the sample and after cooling a glass-like disc is produced that contains the evenly distributed sample which is then analysed directly. A variety of fluxes are used depending on the sample and a wide range of minerals, oxides, slags, cements, etc. can be prepared in this way. The fusion technique is preferred to pelletising because of its lack of particle effects and the fact that standards can easily be made by doping the mixtures with known quantities of the required elements.

Liquids can also be analysed by XRF although most spectroscopists prefer not to put them inside their spectrometers! Liquid samples are usually placed in cups within plastic (usually Mylar) bases. The bases are transparent to X-rays. Only spectrometers with upward-looking X-ray tubes can accept liquid samples and it is usual to perform the analysis in a helium atmosphere. The samples can be solutions, oils containing metallic impurities, etc. The restrictions are that the liquid should have a relatively low vapour pressure and not give off corrosive fumes. Both these

6

problems can lead to damage in the spectrometer.


## 5. INTERPRETATION OF RESULTS

We have seen how XRF works, the range of elements it can cover and the type of samples that can be used. What do we do with the count rate data produced and how does this give us meaningful analytical results?

### 5.1 Inter-Element Effects

Like all measuring instruments an XRF spectrometer needs calibrating since all it produces are a set of numbers related to each measured element. It would be nice if there were a straightforward linear relationship between count rate and concentration. In practice, however, a number of effects conspire to complicate the picture.

Since there is not much point in analysing a completely pure element it can be assumed that every sample contains more than one element. The X-ray absorption and fluorescence effects are not restricted simply to the surface layer of atoms in the sample and occur in the bulk of the sample up to a certain depth depending mainly on the composition of the sample. Thus the secondary X-rays produced by the individual atom interact with other atoms in the sample before finally "escaping" to be measured. The effects can be broken down into three groups: absorption, enhancement and "third-element" effects.

Absorption effects occur when neighbouring atoms absorb the escaping secondary X-rays. This leads to a reduction in the expected fluorescence intensity. The amount of the reduction is dependent on the elements present (since elements have differing absorption characteristics) and on the amount of each element present.

Enhancement effects are produced when radiation from a neighbouring atom excites the analyte atom. This leads to a greater fluorescence intensity than would be expected.

Third element effects are similar to those already described but are related, as the name suggests, to a third element. Thus the radiation produced by a third element causes the second element to fluoresce which in turn causes the analyte to fluoresce.

The combination of all these effects in, for

7

example, an alloy containing, say, a dozen elements can be very complex and lead to measured count rates that are far from the expected.

There are various ways of overcoming these problems. The simplest way is to run standards that are very similar in composition to the samples. A calibration graph of count rate vs concentration can be constructed for the various elements and concentration read off for the count rate obtained from unknown samples.

This method works only over short ranges and providing samples and standards are very similar. By definition the standards used to construct the calibration have varying concentrations of elements and are therefore inherently different from the samples. However this method can be used in production control situations where, for example, alloys are being produced with consistent and well-defined compositions. The method breaks down, however, when new alloys come along or if "off-composition" samples require analysis.

Clearly a method is required that can take account of the inter-element effects and calculate accurate analyses from the raw data. The advent of cheap but powerful computers has enabled this to be routinely performed in the analytical laboratorty.

Over the years much theoretical work has been done on fluorescence intensities and the mathematics of inter-element effects. This has lead to a variety of mathematical methods which can be used to calculate the correction required for a particular element in the presence of others. These methods are beyond the scope of this paper but they are now widely recognised and form a part of most XRF analyses. The calculations involved can become quite complex, especially in the case of sophisticated alloy systems where more than fifteen elements may be present, each affecting the other. In practice many of the effects are small enough to be ignored but some elements, for example chromium in nickel-base superalloys, can easily have eight or ten corrections to be calculated. The introduction of these powerful mathematical methods has meant that, providing all elements producing significant fluorescence are measured, very wide ranging calibrations can be contructed enabling much more accurate analysis of all types of sample.

## 5.2 Precision

Having established an accurate method of calibration we now need to look at the precision of the results.

The processes involved in X-ray emission and fluorescence are, like radioactive decay, random processes. Under constant conditions the same measurement repeated for fixed time periods will lead to different numbers of counts. The emission of X-ray photons follows the laws of probability and must be treated statistically. This randomness introduces a fundamental imprecision into all XRF measurements. It can be shown, however, that if the total number of counts is kept reasonably high, say greater than $10^6$, the coefficient of variation is reduced to an acceptably low level, usually less than 0.1%.

Another source of error is detector dead time. This is due to the fact that after a detector has reacted to an X-ray photon a finite recovery period must elapse before it can react to another. This period can be of the order of several microseconds. If the number of photons reaching the detector is low this will be insignificant but if a high count rate is present the counting loss can be high and can run into tens of per cent. Modern detectors and electronics have reduced this problem but high count rates may still need to be attenuated to avoid large errors.

Other errors may be produced by the spectrometer itself. Temperature fluctuations can cause variation in the d-spacing of the analysing crystal. Instability of the detectors and electronics due to variations in power supply may also produce measurement errors. However most of these random errors have been greatly reduced by improved spectrometer design and they are mostly reduced to very low levels. The remaining errors can usually be corrected for by running monitor samples periodically to measure any instrumental drift.

The errors described so far can be minimised by adopting a suitable counting strategy. Measurements are usually made over a fixed period of time and this can be selected to ensure a level of counts consistent with the precision required. This, of course, must be selected in conjunction with productivity requirements: most laboratory managers would not tolerate an analysis that took half an hour per sample. A modern simultaneous spectrometer can analyse an alloy sample requiring say twenty elements to be measured, in about 45 seconds.

Although this is very rapid a precision, measured as coefficient of variation, of well under 0.5% can often be obtained making XRF a very precise technique.

## 5.3 Useful Range

The useful range of XRF, in concentration terms, is very wide. Traditionally used to measure the higher concentrations of alloy elements from 1% to many tens of per cent the technique is also suitable for measuring large numbers of elements at low levels. In the superalloy industry zirconium is routinely measured at 200 - 300 parts per million and even relatively light elements such as phosphorus and magnesium can be analysed at ppm levels.

## 6. QUALITATIVE ANALYSIS

Although the main use of XRF is for quantification of elements it is also an extremely useful tool for qualitative analysis. By scanning the detector across an angle range and taking measurements at intervals a fluorescence spectrum can be produced showing the various lines produced by the elements. From the Bragg equation the angle at which the line appears can be calculated and the element giving rise to this can be identified. In this way an elemental picture of a completely unknown sample can be built up including rough ratios of the elements.

This is often a very good starting point in the quantitative analysis of a sample since from the data obtained an analytical programme can be decided on.

Examples of this can include analysis of metallic "sludge" from engine oils, identifications of rogue alloys used in incorrect applications and analysis of emission from factory chimneys.

## 7. CONCLUSION

XRF, although relatively expensive, provides a rapid and precise method of analysis for a very wide range of elements and samples. The technique is suitable for metals and non-metals alike and the chemical and physical state of the sample is to a large extent irrelevant. It is non-destructive, i.e. you get the sample back after analysis although it may need to be prepared beforehand. It is suitable for levels from ppm range up to 100% and can be used qualitatively for "fingerprint" identification. It is probably one of the most versatile methods of bulk analysis which accounts for its very widespread use throughout many aspects of the manufacturing industry.

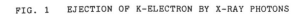

FIG. 1    EJECTION OF K-ELECTRON BY X-RAY PHOTONS

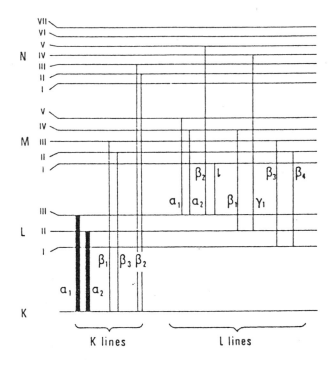

FIG. 2    ENERGY LEVEL DIAGRAM

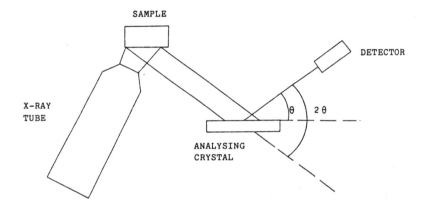

SAMPLE

DETECTOR

X-RAY
TUBE

θ    2 θ

ANALYSING
CRYSTAL

FIG. 3   LAYOUT OF XRF SPECTROMETER

# 2 : Emission and absorption spectroscopy
# J MURPHY

*This paper arrived too late for any corrections to be carried out and is therefore printed as received.*

The author is with Glossop Superalloys Ltd

## INTRODUCTION

Analysis is the measurement of some physical property of a sample  which is specific to that sample and its components.

In emission spectroscopy the property measured is electromagnetic radiation, typically in the wavelength range 150-800nm, this radiation being emitted from a sample which has been provided with sufficient energy to "excite" its component atoms.  Different techniques are available for excitation being generally termed as flames, electric arcs and sparks, plasmas, low pressure discharges, etc.  Each technique produces different excitation characteristics which will influence its eventual analytical performance.  For example, a dc arc emission source provides good analytical sensitivity but poor reproducibility, whereas a spark source offers good reproducibility but poorer sensitivity.

Excitation of a sample, irrespective of the emission source, is not selective enough to allow its component elements to be measured directly.  It is necessary therefore to effect element selectivity through the use of a dispersive optic which produces

14

a spectrum composed of individual element wavelengths. Prisms and diffraction gratings are suitable dispersing optics.

Having isolated a wavelength and identified it as arising from the excitation of a particular element, its detection and measurement is relatively simple: photo-electric detection is most commonly employed in modern instrumentation.

Emission spectroscopy, in its many forms, is one of the most widely used instrumental methods and yet is one of the least well understood due to its spectral com-plexity and somewhat empirical nature. In reporting its applications for high temperature alloy analysis it is essential therefore to establish the basic principles of excitation and discuss various characteristics as they impact on spectral complexity. Section one is given over to this purpose and to discussing the closely allied principles of absorption spectroscopy. Whilst both emission and absorption spectroscopy rely on the same basic principles, absorption spectroscopic methods are much simpler since the influence of spectral complexity is greatly reduced.

The terminology and principles introduced in section one are important to the discussion of section two in which applications and instrumentation are reviewed with particular regard to excitation/absorption sources.

On a note of convention, whilst it is more correct to express wavelengths in terms of nanometers(nm), the common convention used in emission spectroscopy, and this report, is the Angtrom unit (Å). Conversion from Å to nm is however very simple; 10Å is equal to 1nm.


# 1 FUNDAMENTAL PRINCIPLES

## 1.1 EMISSION SPECTROSCOPY

### 1.1.1 Excitation

Atomic spectroscopy, whether emission or absorption, relies on excitation of atoms within a sample being analysed. Whilst it is unnecessary to include a rig-orous quantum mechanical treatment of the process of excitation, it is nonetheless useful to establish the basic principles involved. In this respect it is adequate to use Bohr's atomic model in which electrons

are considered as moving in discrete orbitals around a
central nucleus. Each orbital has an associated energy
level which increases with increasing distance from the
nucleus. Orbitals are defined by means of the principal quantum number (k,l,m,n.... or 1,2,3,4.....). Any
transition of an electron between two quantised energy
level corresponds to the absorption or emission of
energy as electromagnetic radiation, in accordance with
the formula:-

$$E = h\nu = \frac{hc}{\lambda}$$

where h = Planck's constant $(6.624 \times 10^{34}$ Js)
$\nu$ = Frequency of absorbed/emitted light $(s^{-1})$
c = Velocity of light in vacuo $(2.9979 \times 10^8$ ms$^{-1})$
$\lambda$ = Wavelength of absorbed/emitted light (m)
$\Delta E$ = $E_2 - E_1$ where $E_1$ and $E_2$ are energies of the
initial and final states(J)

### 1.1.2 Emission Spectra

Consider an atom in its ground state(ie with its electrons at their lowest energy levels)being supplied with
sufficient energy to promote an electron from a shell
of energy Em to one of higher energy En(see Figure 1).
Such a state is unstable and hence the electron falls
back to its stable state(Em) after a dwell time of
approximately $10^{-8}$s. Electromagnetic radiation is
emitted in the UV/VIS region(1500-8000 A) at a wavelength which is specific to the atom, and hence the
element, which has undergone excitation.
The presence of an element in a sample can therefore be
established through the appearance of a spectral line
at a predefined wavelength. The intensity of this line
is proportional to the number of emitted quanta, and
hence to element concentration.

To conclude the discussion on basic principles at
this stage would be to greatly demean the complexity
of emission spectra for anything but the most simple
hydrogen-like system. It is necessary therefore to
progressively extend the basic ideas to such an extent
whereby it is possible to appreciate some of the problems encountered in using emission spectroscopy for
the quantitative analysis of high temperature alloys.
Additionally, such understanding will explicate the
steps taken to overcome problems inherent in the tech-

nique of optical emission.

### 1.1.3 Multielectron Species

Even simple multielectron species give rise to complcated emission spectra due to electrostatic repulsion between electrons, and the effect of electron spin; both of these phenomena may have a significant impact on orbital energy levels. The result is clearly illustrated in Figure 2 which shows the principal emission series for the common alkali metals, each of which comprises 20 or so lines. Figure 3 is a graphical representation of the emission spectra originating for excitation of sodium. This "term" diagram, often referred to as a Grotian diagram after its creator, illustrates the simpler electronic transitions for sodium. Allowed transitions are shown as diagonals with the emission wavelength included: whether a transition is allowed (ie will in practice produce a spectral line) may only be deduced from quantum mechanical principles, and even then, such calculations may only be applied to relatively simple systems. The most intense lines in any emission spectra are those in which electrons terminate in the ground state. For sodium these are 2853Å, 3302Å 5890Å and 5896Å. Of these, the most intense line is that which originates from the first excitated state which can combine with the ground state in an allowed transition. For sodium, the most intense line, known as the resonance line, is actually a doublet at 5890Å and 5896Å; the well known "sodium D-line".
One remaining piece of information to be gleaned from this term diagram is that an excited electron does not have to return to the ground state via a direct route. Its return may involve several transitions each of which will give rise to the appearance of a spectral line.

### 1.1.4 Shielding and Orbital Degeneracy

Strong nuclear forces are responsible for maintaining electron orbitals; the effectiveness of this attraction diminishes with increasing orbital distance from the nucleus. Electrons are said to be "shielded" by nucleus. Promotion of electrons to a higher energy state is therefore dependent on there being sufficient energy available to overcome the effects of shielding. Accessible energy from most emission spectroscopic sources is normally insufficient to consider electronic transitions occurring in any but the outermost orbital. Transition elements, however, possess several outer orbitals of similar energy(said to be "degenerate")and, furthermore, these outer orbitals are ineffectively shielded by the nucleus. Excitation of transition elements will therefore involve electronic transitions in several orbitals, with a consequential increase in the number of emitted wavelengths. The net result is well illustrated by the example of iron whose emission spectrum contains tens of thousands of wavelengths.

### 1.1.5 Ionisation

Up to now, the discussion has centred on excitation of atoms by the promotion of electrons to higher energy levels. It is necessary however to consider the process of ionisation in which sufficient energy is supplied to an atom to completely remove an electron from its associated orbitals. The positively charged ion thus formed is capable of further excitation, though its emitted wavelengths will be very dissimilar to those of the corresponding neutral atom. Historically, spectra of a neutral atom were described as arc spectra after the manner in which they were produced. Spectra of ionised atoms were similarly classified as spark spectra since they were only capable of being produced by the higher temperature spark discharge.
Since it is now possible to derive arc spectra from a spark source, and vice versa, it is more correct to relate spectral quality to its associated atomic state. Thus, arc spectra are more correctly described as spectra of a neutral atom, and so on.

### 1.1.6 Molecular or Band Spectra

As the name suggests, molecular spectra originate from excitation of molecular as opposed to atomic species. Their distinctive band like appearance, when viewed by spectrometers of low resolution, has resulted in their being referred to as band spectra. Their appearance in emission spectra provides an undesirable source of interference which may render specific spectral regions useless for analytical purposes. (See Figure 4). Emission sources most severely affected are flames and air discharges.

18

### 1.1.7. Wavelength Reference

The complexity of emission spectra may be reinforced byconsidering that there are more than 200,000 lines of neutral and singly-ionised atoms in the wavelength range 2000-8000Å. More than half of these are referenced in wavelength tables as illustrated in Table 1. Roman numerals are used to denote the excitation state of the emitting atom. (I) relates to a neutral atom and (II) to a singly ionised atom. No Roman numeral means that the origin is unknown.

### 1.1.8 Spectral Complexity and Analytical Sensitivity

At any instant in time the amount of energy accessible for excitation is constant for any particular technique. It follows, therefore, that the greater the number of excitation states the less is the available energy per excitation state, the lower will be the analytical sensitivity. Intuitively it can be surmised that element sensitivity varies inversely with the number of excitation states produced. A good illustration of this point is provided by comparing the sensitivity of arc and spark emission spectroscopy. Energies associated with spark discharges are far greater than those of arc discharges and hence spark spectra contain far more wavelengths than corresponding arc spectra. Consequentially spark emission techniques are less sensitive than are arc techniques as will be developed in subsequent discussion on applications.

### 1.1.9 Self Absorption or Reversal

In emission spectroscopy the relationship between intensity and concentration is rarely linear over a wide concentration range. Above a certain level, the rate of change in intensity decreases with increasing concentration. The phenonemon largely responsible for such curvature is self absorption, also known as self reversal. Self absorption may be explained by considering the effect of higher populations of ground state atoms(vapourised but unexcited) present in a discharge with increasing concentration. Whilst there is a similar increase in excited atoms, the overall light yield is reduced due to absorption of emitted radiation

by the surrounding ground state atoms. Self absorption
is very pronounced in dc arc techniques in which there
is a high rate of sample vaporization. Lines of neutral atoms and resonance lines in particular, tend to
be affected to a greater extent than corresponding
ionic lines. Self absorption is undesirable since it
results in deteriorating precision with increasing
curvature.

### 1.1.10 Line Width

All spectral lines have finite natural widths since
electron energy levels are not "sharp". Natural widths
are of the order of $10^{-4}$ Å. Lines may however be considerably broadened by a number of phenomena which are
dependent upon prevailing excitation conditions:-

    i Doppler broadening is a temperature related
      phenomenon such that the higher the temper-
      ature, the greater is the effect. High
      plasma temperatures may broaden lines by as
      much as 0.1-0.2Å.

    ii Pressure broadening. High pressure discharges
      may result in broadening which is similar in
      magnitude to that caused by the Doppler effect.

    iii Self absorption may have a similar influence on
      the width of a spectral line.

    In cases where spectral interference, due to line
overlap, is a problem, the position may be made much
worse through the contribution of line broadening by
one or more of the above effects.

### 1.2 ABSORPTION SPECTROSCOPY

### 1.2.1 Basic Principles

Discussion and description of absorption spectroscopy
will be limited to considering atomic absorption
spectrophotometry(AAS). Since the same basic principles of excitation apply equally well to both absorption and emission spectroscopy, it is only necessary
to highlight areas where the techniques diverge.

    In AAS, radiation from an external light source
(eg hollow cathode or electrodeless discharge lamp),
whose energy corresponds to that required to produce
an electronic transition from the ground state to an

excited state, is directed through an "atom cell".
The function of this "atom cell", either a flame or
electrothermal device, it to produce a population of
ground state(free, unexcited)atoms of the element
whose concentration is to be determined.  These ground
state atoms absorb  a proportion of incoming radiation
from the source where it corresponds exactly to the
energy required to produce a transition of the test
element from ground state to an excited state.
Unabsorbed radiation is then passed through a monochromator, which isolates the exciting spectral line,
and onto a detector.

The radiation absorbed from the incoming light is
proportional to the population of absorbing ground
state atoms and hence the concentration.  Absorption is
measured by the difference in transmitted signal in the
presence and absence of the test element.

The fundamental law underlying absorption spectroscopy is Beer's Law.  It states that when a plane parallel beam of monochromatic light enters an absorbing
medium at right angles to the plane parallel surfaces
of the medium, the rate of decrease in radiant power
with the length of light path b, or with concentration
of absorbing material c, will follow an exponential
progression:-

$$T = \frac{p}{Po} \quad \text{or} \quad A = \log_{10} \frac{Po}{p}$$

where T = transmittance
      A = absorbance
     Po = intensity of incident beam
      p = intensity of transmitted beam.

## 1.2.2 Spectral Interferences

In sharp contrast to emission spectroscopy, absorption
rarely suffers from problems related to line overlap.
Selectivity is realised by the use of an external light
source which is chosen to emit radiation characteristic
of the element to be determined.  The major source of
spectral interference is that of light scattering which
is more commonly encountered in using electrothermal
atomisation devices.  Particles of ash, smoke and salts
thrown into the light beam during atomisation cause the
incident light to be scattered thereby decreasing the
transmitted intensity.

### 1.2.3 Self Absorption

Self absorption can be a major problem in atomic absorption since measurements are commonly made using resonance lines. As with emission spectroscopy. self absorption increases with increasing concentration.

## 2    APPLICATIONS AND INSTRUMENTATION

### 2.1 INTRODUCTION

The suitability of a technique for any particular application is dependent upon a number of important, though often incompatible, factors.  Essentially, these factors may be stated as follows:-
1.   The technique should be capable of analysing a wide range of elements and yet provide a high degree of selectivity for elements of interest.
2.   It should be capable of providing the required sensitivity and precision.
3.   In order to maintain long-term sensitivity and precision, instrumentation must be stable.
4. Interferences should be minimal and memory effects negligible.
5.   Capital and operating costs should not be excessive
6.   Convenience of operation may be an important consideration.
7.   Speed of operation may be critical, especially within a quality control environment.
8. Flexibility to analyse a variety of sample shapes, sizes and forms may be of importance.
9. Sample preparation should ideally be rapid, simple and reproducible.
10.The technique should be reliable and preferably of proven design.
11. Adequate support, both technical and service, should be provided by the instrument manufacturer.
    Obviously, the relative merits of these desirable properties will depend on an organisation's particular analytical requirements, and on the size and nature of the organisation.  Mutual exclusivity of some of the properties will, however, invariably lead to instrument choice being a matter of compromise.
    Exclusion of accuracy from the preceding list is· neither an oversight, nor is it to suggest that accur-

acy is unimportant. It is omitted so as to highlight
the fact that accuracy may be adversely affected by
factors other than instrument performance. Sample
heterogeneity and/or metallurgical condition may have
a significant influence on accuracy, particularly in
the case of direct methods(ie analysis of a solid
sample) applied to complex alloy systems.

Accuracy is also influenced by the quality of reference materials used to establish an instrument calibration. It is worthwhile to note that though a reasonable range of certified reference materials(discs and
millings)is available for ferrous-base materials, the
same is not true of nickel, cobalt and titanium alloys.
In these applications, calibration is effected through
the use of "in-house" standards, which may lead to considerable inaccuracy if suitable precautions are not
employed to verify calibration accuracy. Solution
methods are largely unaffected by the availability of
reference materials, for calibration purposes, since it
is possible to employ a totally synthetic calibration
route using pure materials.

The following discussion on applications of emission and absorption spectroscopy for high temperature
alloys analysis largely follows an historical path from
past to present. Such an approach is useful in providing a view of the natural progression which has taken
place both in method applications and in developments
of instrumentation. Advantages and disadvantages of
instrumentation and techniques are stated to provide an
overall picture of emission and absorption spectroscopy
and to help to illustrate the necessity for developments in both areas. Competing techniques such as X-
ray spectrometry are included where relevant comparisons are applicable. A comparative summary of various
methods is provided in Table 2, and data is also presented for individual techniques(Tables 3 to 12). It
should be understood that the results presented are
neither the best achievable nor the worst obtained.
They are an overall picture of the type of performance
which may be expected, though this of course will depend on a large number of factors such as individual
instrumentation, working environment, an organisation's
requirements, etc.

Concentration Convention.

An arbitrary convention has been adopted for the purpose of this report to avoid any misunderstanding of

23

the terms "ppm" and "$\mu$g ml$^{-1}$". Thus, within the context of this report ppm refers to weight percentage and not $\mu$g ml$^{-1}$.

Thus, .001 wt% = 10ppm


## 2.2 FLAME EMISSION SPECTROSCOPY (FES)

Flame emission spectroscopy may be considered to be of historical interest only, having been largely superceded by other emission techniques and by atomic absorption spectroscopy(AAS). It is the oldest and simplest spectroscopic method and its application for the analysis of ferrous-base alloys has been widely reported and reviewed (1), (2).

In operation, a solution of the sample to be analysed is sprayed into a flame, either directly via a total consumptive burner, or indirectly after passage through a spray chamber atomiser. Flame temperatures of 1900-3000°C are achievable depending on the choice of fuel-oxidant. The flame serves to evaporate solvent dissociate molecules into atoms, and excite the atoms thus generated.

Major advantages are those of low cost instrumentation and the ability to employ a synthetic calibration route.

FES has many disadvantages. Being a solution technique, it is slow because of the need to incorporate a dissolution stage. Additionally, though sensitivities are generally good, it is necessary to allow for dilution effects of the solvent. Factors such as solvent purity and viscosity may further disadvantage the method. Band spectra(1.1.6) caused by flame combustion processes render portions of the spectral range unusable. Inferior optical dispersion of low cost instrumentation leads to severe interference due to line overlap. Stable compounds with high dissociation energies may be found in the flame e.g $SiO_2$, $Nb_2O_5$. Detection limits of such elements by FES are correspondingly high.

Calibrations are non-linear with deviations at both high and low concentrations. At high concentrations self absorption(1.1.9) and viscosity effects, give rise to loss in intensity resulting in negative curvature. The severity of this curvature is sufficient to render flame emission useless for analysis

24

of high concentrations. At very low concentrations,
ionisation(1.1.5) promotes positive curvature, though
this may be overcome through the addition of a suitable
ionisation buffer.

Though AAS has largely replaced FES, there are
certain applications where FES outperforms flame AAS
(3) and hence most atomic absorption spectrophoto-
meters are capable of being operated in an emission
mode. Detection limits and analytical precisions
extracted from a number of published papers are
summarised in Table 3.

### 2.3 D.C. ARC EMISSION SPECTROSCOPY

Conventionally, a dc arc source consists of a regulated
power supply which furnishes 110-220v at 3-30A, current
being adjustable by a variable resistor. A dc arc con-
stitutes a stable electrical discharge of high current
density; essentially a plasma, which like all plasmas
is a complex phenomenon. High plasma energy in the
cathode region produces temperatures of about 5000°C
which results in thermal emission of electrons from
the cathode. These electrons, accelerated towards
the anode, produce anode temperatures of approximately
6000°C through a collision mechanism. Hence, a sample
may either be anodic or cathodic depending on the volatility of components to be measured; samples which are
difficult to vapourise are often made anodic. A graphite rod machined to a conical point is commonly used as
the counter electrode. Techniques are therefore described as "point to plane" where "point" refers to the
counter electrode and "plane" defines the sample form
as being "flat".

Once an arc has been initiated, the sample is
rapidly vaporised into the plasma where it is excited.
The resulting emission spectrum, in contrast to that
derived from a spark source, exhibits low background
intensity and fewer spectral lines which are predomin-
ately those of neutral atoms(see 1.1.5). Inherently,
the dc arc source provides good sensitivity with detection limits of the order 0.1 - 1ppm being typically
achieved. The high sample vaporization rate responsible for such good sensitivity, leads however to high
plasma density, which in turn causes pronounced self
absorption. DC arc sources are therefore unsuitable
for applications requiring quantitative analysis at

alloying concentrations. Quantitative analysis is further restricted by poor analytical precision arising from arc "wander" across the sample surface. Relative precisions are typically of the order of 10-30%.

Quantitative analysis also requires using solid metal reference materials whose composition is similar to the sample being analysed (matrix-matched) since the excitation characteristics of a dc arc are strongly influenced by matrix excitation.

At this stage, it is appropriate to discuss instrumentation with which the conventional dc arc source was historically associated. Rather than review all possible source/spectrograph combinations it is advantageous to concentrate on one specific design since the same basic operating principles apply to most of the early emission spectrographs. The Littrow prism spectrograph, illustrated in Fig 5, is chosen because of its popularity in the UK. In common with most emission spectrographs of its time, the Littrow instrument employed a quartz prism for wavelength dispersion, the resulting spectra being recorded on a photographic plate. The plate, after suitable processing, could be examined qualitatively using a plate viewer or quantitatively using a scanning microphotometer. Wavelength identification was achieved by generating a spectrum of iron on each exposed plate which was compared to a reference spectrum of iron. When using the instrument quantitatively it was necessary to compensate for source instability and variations in emulsion characteristics of the plate by ratioing each wavelength of interest to an homologous reference line. Ideally, this reference line had to possess similar excitation characteristics to the measured line, be of constant or known composition and have a wavelength as close as possible to the measured line. Much of the early success of dc arc emission may be attributed to the fact that it removed the need for a time consuming and often difficult sample dissolution stage.

Accepting that dc arc techniques were only suitable for trace element analysis, the major disadvantage was in obtaining sufficient reliably analysed samples with which to effect instrument calibration. Such samples were difficult to produce and standardise and, because of the requirement for matrix-matching, each alloy type required its own range of matched calibration samples. A number of chemical separation pre-concentration methods(4),(5) were developed which utilised synthetic calibration routes, thereby obviating the need for alloy reference samples. Additional benefits of these and other methods(6) (7) were in improving detection limits of certain deleterious trace elements. Typical detection limits and precisions obtained by some of these methods are presented in Table 4. Obvious disadvantages of such methods were that they were time consuming and often required a high degree of chemical skill.

DC arc emission spectroscopy has largely disappeared from the quality control type environment in which it was originally used, having been replaced by alternative techniques which will be discussed later. It is however one of the few remaining techniques which allows a spectrum to be viewed and is therefore still to be found in many laboratories, being mainly used for qualitative investigations.

One important application in which dc arc emission spectroscopy has flourished in recent years is in portable spectrometers used for scrap sorting and product verification. In these instruments the source is incorporated into a hand held pistol which is presented to the material being examined. A fibre optic, connected to the pistol and housed in a protective sheath, provides the means of light transfer from pistol to spectrometer. Dispersive optics and signal processing is housed in a compact rugged, dust-proof cabinet. Relative precisions of 10-30% mean that the spectrometer is only suitable for spectral "fingerprinting", though it has been highly successful for the majority of scrap sorting applications. Its robust design makes it particularly suitable for foundry or melting shop environments. Argon flushed pistols may be used in certain circumstances to provide performance which is comparable to that of a larger laboratory type instrument. Transmission properties of the glass fibre light guide do not allow carbon to be analysed, at this moment in time.

2.4 SPARK EMISSION SPECTROSCOPY

Spark discharges are short-duration dc or ac arcs whose spectral and performance characteristics are in

27

total opposition to continuous arcs. In a spark discharge the power conversion per cycle is much greater than in continuous arcs, and hence the instantaneous temperature in the spark gap is correspondingly higher. Thus, spectra generated in a spark discharge are complex, predominantly exhibiting lines of ionised atoms but with a reasonably high population of neutral atom lines.
Background intensities are higher than in continuous arcs. Sample vaporization rates in spark discharges are much lower than in continuous arcs, which leads to lower plasma density and hence markedly less self absorption. With regard to analytical performance, poor sensitivity results in limits of detection of spark methods being relatively poor, though this is compensated for by analytical precision which is very good. The combination of good precision and low self absorption results in spark emission methods being suitable for quantitative analysis of relatively high concentrations.

At the outset, it is unnecessary to describe the "myriad" of commercially available spark sources with names such as; monoalternance, intermittent arc, condensed arc, high repetition rate source, high energy pre-spark, unisource, etc. It is however important to comment on some of the more general properties of these types of spark sources in their effect on spectral characteristics. Firstly, the power conversion per cycle(intermittent energy) determines the amount of sample vapourised and is therefore influential on spectral intensity. Secondly, the duration of a spark (discharge frequency) determines the type of excitation which predominates. Thus, for a constant energy, the shorter the discharge frequency the more predominant are neutral atom lines.
It is therefore possible, through careful selection of certain source parameters to confer,on a spectra, qualities to suit a particular application. Furthermore, it is often possible to select different source conditions for pre-spark and integration stages.
A high energy pre-spark may be necessary for difficult inhomogeneous materials. Homogeneity is achieved through micro-melting of the sample prior to signal integration which is often at a lower energy. A computer-controlled variable frequency source is also

available which permits discharge duration to be
varied during the integration stages. This provides
a means of optimising integration conditions on an
element by element basis.

Before reviewing specific applications of spark
emission spectroscopy it is pertinent to examine
instrumentation used. Rather than scrutinise individual stages of instrument development, it is prudent
to discuss modern spectrometers and their development,
in comparison with the early types of emission spectrographs already described. An elementary outline of
modern types of instrumentation is also included.
Almost all modern spectrometers use reflective
gratings for wavelength dispersion. Gratings are preferred to prisms since they provide much higher resolution, greater dispersion and a dispersion which is
almost linear with wavelength.
The wavelength range of early spectrographs(2000-
8000Å) meant that certain elements eg C,S,P,B,As,Se,
whose most useful wavelengths were in the range 1600-
2000Å) could not be accessed. The 50ppm lower reporting limit for As(2350Å) quoted in Table 4 is a reflection of this inaccessibility, the most sensitive As
wavelength being 1973Å. In modern instruments it is
possible to access the range 1600-2000Å by excluding
air from the spectrometer either through a vacuum
spectrometer or by purging with a gas such as nitrogen
which is transparent to wavelengths in the 1600-2000Å
range.
Though early spectrographs enabled simultaneous
recording of a spectrum, subsequent translation into
concentration was a slow and tedious process. In modern
emission spectrometers the use of photo-electric
detection and computerised data processing has drama-
tically improved both speed and ease of operation.
Modern simultaneous spectrometers are capable of analysing up to 60 elements in a sample in less than one
minute. Whilst computerisation offers many additional
benefits it is important to remember that the heart of
a spectrometer is its source and optical system. Thus
it is better not to have an interference than to have
to correct for it!
Most of the larger modern emission spectrometers
utilise reflecting concave gratings in a Paschen Runge
mounting based on the Rowland circle concept. In this

arrangement(see Figures 6 & 7), entrance slit, grating
and exit slits are configured to lie on the circumfer-
ence of a circle whose diameter corresponds to the
radius of curvature of the grating. Light from an
emitting source enters the spectrometer through a primary (entrance) slit. It is directed onto a grating
where it is dispersed into its component wavelengths
which are reflected to the detector system. In simultaneous instruments a number of fixed secondary(exit)
slits are  positioned around the Rowland circle at pre-
determined wavelengths. Light passing through these
slits is directed onto photomultiplier tubes using
torroidal mirrors(Figure 7) or via light guides(Figure
6). In sequential spectrometers there is, essentially,
one single secondary slit and photomultiplier assembly
which can be driven to any selected wavelength.
Sequential spectrometers are more flexible than simultaneous spectrometers but slower. The ultimate in
speed with flexibility is provided by a combined simul-
taneous/sequential spectrometer which incorporates twin
excitation stands(see Figure 6).

Having already stated that spectrometer choice is
largely a matter of compromise, it is useful to report
one example which illustrates this point quite clearly
Possibly the greatest compromise made is that of wavelength selection, particularly where an analytical
application is complex. Table 4 illustrates such a
programme for a twin stand spectrometer(argon spark
plus hollow cathode sources) capable of analysing major,
trace and low boiling point elements in Fe, Ni, Co-base
alloys. Of the 44 listed wavelengths, 29 are in the
range 1500-3000Å with only 15 in the range 3000-6000Å.
Limitations of the spectrometer, in respect of space
available for sighting of secondary slits and photomultiplier tubes, resulted in having to measure some
wavelengths in the second order diffraction. Furthermore it was necessary to sacrifice some preferred wavelengths to accommodate all elements in the programme.
For example, the most useful lines for As and Se at
1972Å and 1961Å respectively, could not be accommodated
in the same programme, hence As 2350Å was used as a
compromise.

One solution to the problem of insufficient space
is to use a split optical configuration(Figure 6) in
which the primary grating is used for wavelengths in
the in the range 1600-4850Å and the secondary grating

30

is used for higher wavelengths in the range 5300-8000Å
Another rather elegant solution is to use several
optics each with its own grating, slits and photomultipliers. Light transfer from the source to each
optical module is via fibre optic cables arranged
radially around the source. Though fibre optics will
not transmit wavelengths in the vacuum/UV region, their
measurement is made possible by incorporating a small,
conventionally mounted, vacuum polychromator. The role
of fibre optics will undoubtedly have a great influence
in future spectrometer developments.

Applications of spark emission spectroscopy for the
analysis of high temperature alloys are best considered
by comparing its performance in this area to that of
"competing" techniques such as X-ray spectroscopy and
combustion methods for carbon and sulphur.

At low concentrations, spark emission is characterised by good precision which deteriorates with
increasing concentration. The reverse applies to X-ray
spectroscopy which offers exceptionally good precision
at high concentrations and poorer precision at low
copncentrations.

Both techniques may be used to analyse a wide range
of elements though X-ray spectroscopy is effectively limited to elements of atomic number $>$10. It
remains to be seen whether recent developments in thin
wafer crystals will allow X-ray spectroscopy to be used
for boron and carbon analysis. Spark emission is routinely used for carbon, boron and nitrogen though its
performance for these elements is inferior to combustion methods.

The linear relationship between intensity and concentration is a noteable feature of X-ray spectroscopy.
Spark emission calibrations, by comparison, are rarely
linear over a wide concentration range, often necessi-
tating two and even three wavelengths for use at different concentrations.

Interference corrections in spark emission are
largely empirical, hence the best performance is
obtained by using matched standards and samples. Interferences in X-ray spectroscopy, particularly absorption
and enhancement, are more predictable and can be expressed mathematically thereby alleviating the need to
use matrix-matched standards.

Spark emission, unlike X-ray spectroscopy, is a

destructive technique in which standards are consumed
and will therefore require replacing.

Actual performance of spark emission and X-ray
spectroscopy is presented in Tables 6-8 in accordance
with the following discussion.  Spark emission data was
generated on a ARL B34000 simultaneous spctrometer
whose wavelength programme is presented in Table 5.
Calibration of the spectrometer for the sample whose
results are presented was by means of matched standards.  X-ray data were generated on the same sample
using a Philips PW1400 sequential X-ray spectrometer
which employs a multi-matrix calibration model.  It
should be noted that whilst the PW1400 is very much
"state-of-the-art", the B34000 is some 6 years older.
Additionally, the B34000 is a twin stand configuration
with the spark stand being given "lower priority" in
respect of performance requirements.

Lower reporting limits, for elements of interest
in high temperature alloys, are quoted for both instruments in Table 6.  These limits are, in most cases, at
least three times greater than measured detection
limits to allow for zero-point calibration error and
instrument drift.  Whilst spark emission is shown to
provide better reporting limits than X-ray, it is perhaps debatable as to whether such improvement is necessary for routine quality control applications.

Table 7 shows the short term precision calculated
from 10 replicate analysis on a typical nickel-base
casting alloy.  It can be seen that spark emission
offers better precision than X-ray at "lower" concen-
rations(Zr,Fe).  Precisions of the two techniques are
comparable at "medium" concentrations (Ti,Ta,Hf) but
X-ray is superior at concentrations greater than 2.5%.
The superiority of combustion techniques for carbon is
also shown, data having been generated on turnings
analysed using a Leco CS244 instrument.

The effect of instrumental drift on long term precision was assessed by analysing the sample once per
week for a period of eight weeks.  Normal precautions
were used to compensate for drift such as using drift
correction samples, etc.  Data presented in Table 8
when compared to that in Table 7 shows, as expected,
that long term precision is worse than short term
precision for both techniques.  The deterioration is
however far more pronounced for spark emission.

Demanding alloy specifications imposed on the

superalloy industry mean that spark emission cannot
be considered for the analysis of high concentrations.
Spark emission is however commonly used in the steel
industry where its superior element coverage and generally lower cost is a definite advantage.  Typical
performance on a highly-alloyed steel sample is presented in Table 9.


### 2.5 LOW PRESSURE DISCHARGES

The two techniques which need to be considered in this
category are hollow cathode and glow discharge.  Since,
however, the technique of hollow cathode emission
spectroscopy(HCES) is discussed at length in an included report on "Gases and Trace Elements", its discussion here will be summarily treated.

### 2.5.1 Hollow Cathode Emission

The HC source of Thornton(8) is used, in combination
with conventional emission spectroscopy. for the
simultaneous analysis of fourteen low boiling point
elements whose presence in high temperature alloys has
been shown to have a largely deleterious influence on
mechanical properties.  Elements and their concentration ranges are quoted in Table 10.
      In operation, HCES uses approximately 10mg of
sample(millings) placed in a hollow graphite cathode
which is fixed onto an electrode holder and inserted
into the hollow cathode lamp.  This lamp in the form
of a stainless steel body is evacuated and then continuously flushed with helium at low pressure, typically
10-30 mbar.  The lamp operates at a potential of 400-
500v, power dissipated in the lamp being linearly
ramped through the integration period.  Increasing
cathode temperature causes selective volatilisation
of elements in the sample according to their boiling
points;  excitation of these elements occurs in the
cathode region.  Spectral lines of the low boiling
point elements are intense and narrow due to the lack
of Doppler and pressure broadening(1.1.10).  Excitation of the matrix, which would otherwise result in a
complex emission, is avoided by limiting the maximum
power dissipated in the lamp.  Analysis time is
approximately 7 minutes per sample.
      HCES, in this application, provides much better

sensitivity than the dc arc emission techniques and is
much quicker than AAS.  It does nowever require
metallic standards and analytical precision is not
good(5-15% relative).

### 2.5.2  Glow Discharge Emission

The glow discharge lamp(GDL) according to Grimm(9)(10)
is shown schematically in Figure 8.  The anode and
water cooled cathode are precision machined blocks
which form the sides of a discharge chamber, anode
and cathode being separated by a suitable insulator.
One end of the lamp is closed by a quartz window,
through which the discharge is transmitted, the other
end by a flat conducting sample.  The lamp is evacuated
and then continuously flushed with argon at low pres-
sure(1 to 20 mbar).  The discharge is sustained at
voltages in the range 400-1800v with currents of 100-
300 mA.  Argon ions, formed in the electric field of
the lamp continuously bombard the cathodic sample
causing atoms to be ejected from the sample sur-
face(cathode sputtering).  These atoms are subsequently excited in the anode region through a collision
mechanism.
    Glow discharge is characterised by an absence of
self absorption, which provides linear calibration
curves over several orders of magnitude of concentration.  Unlike other emission techniques, excitation
does not involve any thermal processes, sputtering of
the sample surface is uniform and the excitation is
independent of electrical or thermal properties of
the sample.  In practical terms, this results in low
matrix interferences and extremely good analytical
precision which approaches that of X-ray spectroscopy.
Detection limits for most elements are slightly
inferior to spark emission.  Since the glow discharge
is a low temperature, low pressure process spectra are
sharp which reduces line overlap interferences.
    In considering the advantages of glow discharge
it is difficult to understand why it is not more widely
used, particularly in the field of high temperature
alloy analysis as a "competitor" to X-ray spectroscopy.
Its analytical performance for the analysis of two
steel samples is presented in Table 11.
    It can also be used for non-conductive materials
by pelletising the sample with conductor such as

34

copper. Uniform sample sputtering means that the
technique can be successfully used for surface analysis
and depth profiling applications.

### 2.6 ATOMIC ABSORPTION SPECTROSCOPY

Since its introduction by Walsh in the mid 1950's,
AAS has proved itself to be the most powerful analytical technique ever presented for the analysis of
solutions. Whilst in certain quarters it is suggested
that plasma emission techniques will eventually
replace AAS, this seems very doubtful. It is important
to understand that the two techniques are complementary
rather than competitive since both have strengths and
weaknesses. AAS certainly provides a unique combination of simple operation, low cost instrumentation,
good precision and good sensitivity, which make it an
invaluable addition to any laboratory. Furthermore,
calibration can be effected by a number of routes
ranging from totally synthetic, standard additions
on certified reference materials, whichever is most
appropriate. The fact that analysis of metal by AAS
requires a time consuming sample dissolution stage,
coupled with its present inability to perform simultaneous analysis means that it cannot compete against
spark emission and X-ray spectroscopy where time is an
important consideration. It is however a good
"back-up" instrument for verifying the accuracy of
results obtained from such instrumentation. AAS
has in recent years grown in stature to the extent
that it is now accepted internationally as a standard method of analysis for a number of elements
(11) (12).
    The basic components of an atomic absorption
spectrophotometer are schematically illustrated in
Figure 9. Whilst modern instrumentation is heavily
influenced by microprocessors and computerisation,
the optical design has largely remained unchanged
from that of the early instruments. AAS revolves
around the "atom cell" which in the case of flame
AAS comprises the burner-nebuliser system. The "pre-
mixed" or "laminar-flow burner" is commonly used in
modern instrumentation and will be discussed.
    A solution of the sample is sucked up a capil-
lary tube through the action of a partial vacuum
caused by gas flowing across the end of a venturi.

The aerosol thus produced is mixed with fuel and
oxidant and passes to the burner. Baffles inside the
mixing chamber ensure thorough mixing and serve to
prevent larger droplets reaching the flame. Approximately 85% of the sample is discarded in this way.
A 10cm burner is usually employed for air-acetylene
operation, this being reduced to 5cm for nitrous oxide/
acetylene. Temperatures of these two flames are
approximately 2300°C and 3080°C respectively. The
higher temperature capability of nitrous oxide/
acetylene is essential for elements which form stable
compounds in the flame such as oxides of aluminium,
tantalum, niobium, zirconium and hafnium. It is however much "noisier" than air/acetylene and "sooting"
leads to burner blockage.

A number of different types of non-flame atomisers
are available to provide improved sensitivity where
required. Most of these utilise some form of resistance heated rod or furnace which is protected from
oxidation by being bathed in an inert gas such as
argon. Generally they are known as "electrothermal
atomisers". The vast improvement in sensitivity
provided by such devices arises from increased dwell
time of ground state atoms in the optical path when
compared to flames. Such systems also offer better
control of the chemical environment through the
inclusion of an inert atmosphere. Disadvantages of
non-flame atomisers are considerable. Sensitivities
of elements which form stable carbides(eg Si, Ti, Ta,
Nb, etc.) are poor. Methodology utilising non-flame
techniques is not readily transferable between instruments. Background correction is essential since non-
specific absorption interferences(1.2.2) are high.

The use of non-flame AAS for the analysis of
deleterious trace elements in nickel and cobalt-base
alloys has been widely reported(13),(14),(15). Its
performance in this area of analysis is reviewed in an
included paper entitled "Gases and Trace Elements".

Applications of flame AAS for the analysis of high
temperature alloys have been widely reported and reviewed (16),(17). A number of composite schemes have
been developed for Ti alloys(18), steels(19),(20),(21)
and Co alloys(22). All methods reported relative
precisions of 1 percent or better for minor and major
constituents with element concentrations up to 20%.
Method accuracies, based on analysis of included

certified reference materials, were on the whole good, even at high concentrations. All methods employed a synthetic calibration route in which the alloy base of the standards was approximately matched to that of samples analysed.

With the advent of computer controlled simultaneous flame AAS, and utilisation of rapid dissolution methods, or direct solids nebulisation, it will be interesting to re-examine the position of flame AAS in two years or even less. Flame and non-flame detection limits are quoted in Table 12.

2.7 PLASMA EMISSION DISCHARGES

The four main types of plasma discharge source are; direct current plasma(DCP), microwave induced plasma (MIP), capacitively coupled microwave plasma(CMP) and inductively coupled plasma(ICP). Of these four, inductively coupled plasma(ICP) is most widely used and thus discussion of plasma discharges will be limited to ICP.

Whilst the advantages of ICP as an emission source were recognised in the mid 1960's (23),(24), almost 10 years elapsed before its dramatic entry into the analytical arena. During its relatively short lifetime it has made a considerable impact on the field of atomic emission spectroscopy, not only in its own right, but also in re-focusing interest in the general area of emission spectroscopy.

The basic operating principles may be explained by reference to Figure 10 which illustrates an ICP torch. Energy, supplied to the water cooled coil by a radiofrequency generator producing a forward power of up to 3kW at between 5 and 30 MHz, is coupled to argon gas flowing through the outer tube  of a concentric three-tube quartz torch. The plasma is initiated by seeding with electrons after which time it is self-sustaining through a process of collision and ionisation.

Temperatures attained in the plasma are of the order of 10,000K though the analytical region, in the tailflame, is only 5000-6000K. A sample in the form of an aerosol is introduced into the torch via argon carrier gas which punches a hole in the base of the plasma.

It has long been recognised that sample transport into the plasma provides its major drawback, due to the low transport efficiency of pneumatic nebulisers (typically only 3% of the sample reaches the plasma).

37

High solid concentrations result in the additional
problem of burner blockage. Considerable effort has
been expended in attempts to improve nebuliser design
and to a large extent these have been successful.
Nebulisers are now available which provide much higher
transport efficiencies and may be used for concentrated
solutions and even slurries.

ICP provides good limits of detection its sensitivity being derived from the high plasma excitation
temperature. The high temperature is also sufficient
to ensure that chemical interferences are minimal.
The area of application where these factors are particularly important is that of refractory element analysis (eg V,Mo,Zr,B,Ta,Hf,Nb) whose sensitivities by AAS
are particularly poor. Detection limits for these and
other elements of interest in high temperature alloys
analysis are quoted in Table 12.

The "optical thinness" of the ICP viewing zone
means that self-absorption is extremely low and hence
calibrations are linear over a wide dynamic concentration range.

Analytical precision is of the order of 1% relative, though significant improvements may be realised
by careful choice of internal standards.

As an emission source, ICP may be configured with
any type of optical emission spectrometer to provide,
in the case of simultaneous instrumentation, a rapid
method of analysis or in the case of sequential
spectrometers, a high degree of flexibility. It is
however necessary to allow for inherent drawbacks of
the ICP source: its high excitation temperature produces complex spectra which may result in serious line
overlap interferences. The effects of line overlap
are compounded by significant Doppler and pressure
broadening of the spectral lines. One rather interesting solution to the problem of resolution is provided by the echelle spectrometer (see Figure 11) which
provides the ultimate in wavelength dispersion whilst
retaining compactness.

CONCLUDING REMARKS

This brief review of emission and absorption spectroscopy is by no means complete, though it is hoped that
the reader has gained a valuable insight into principles of operation, terminology, instrumentation used,

the types of performance capable of being provided,
and some of the problems which may be encountered.
There are a number of good and interesting textbooks
(26), (27) which can be referenced to provide a more
detailed treatment of individual areas of technique
and instrumental considerations.

ACKNOWLEDGEMENTS

Firstly I would like to thank Glossop Superalloys Ltd.
for their support in preparing this publication and
for allowing the inclusion of data generated in their
laboratory.
Secondly, I gratefully acknowledge the assistance of
a number of companies in providing both data and dia-
grame, and for allowing their publication.  In particular my thanks to to Applied Research Laboratories
Ltd., Philips Scientific and Hilger Analytical.
My thanks are finally extended to Addison-Wesley
Publishing Company for allowing me to reproduce
certain material from their publication Chemical
Instrumentation (26).

REFERENCES

1.  V.A. Fassel, R.W. Slack, R.N. Kniseley:
    Anal.Chem. 1971, 43,(2), 186.
2.  R. Mavrodineanu: "Bibliography on Flame Spectroscopy.
    Analytical Applications: 1800-1966" National Bureau
    Standard(US) Misc. Publ. 281, 1967.
3.  M.E. Hofton, D.P. Hubbard, F. Vernon: Anal.Chim.Acta.
    1971, 55, 367
4.  B.E.Balfour, D. Jukes, K. Thornton: Appl.Spec.1966
    20,3,168.
5.  G.S. Golden, M.G. Atwell: Appl.Spec. 1970, 24,5,514.
6.  M.G. Atwell, G.S. Golden: Appl.Spec. 1970, 24,3,362.
7. M.G. Atwell, G.S. Golden: Appl. Spec. 1973, 27,6,464
8.  K. Thornton: Analyst 1969, 94,958.
9.  W. Grimm: Naturwiss  1967, 54,586.
10. W. Grimm: Spectrochim. Acta. 1968, 23B, 443.
11. BS 6783: Part 3: 1986.  ISO 7520-1985
12. BS 6783: Part 1: 1986.  ISO 6351-1985
13. G.G. Welcher, O.H. Kriege, J.Y. Marks: Anal.Chem. 1974
    (46), 9, 1227
14. J.E. Forrester, V. Lehecka, J.R.Johnston, W.L. Ott:
    Atom. Abs. Newsletter 1979, 18, 4, 73

15. S. Backman, R.W. Karlsson: Analyst, 1979, 104, 1017
16. P.H. Scholes: Analyst 1968, 93, 1105, 197
17. T.S. Harrison, W.W. Foster, W.D. Cobb: Metallurgia and Metal Forming(Nov) 1973, 361.
18. D. Myers,: At. Abs.Newsletter, 1967, 6, 4, 89
19. J. Husler:,At. Abs. Newsletter, 1971,10,2, 60
20. W.R. Nall, D. Brumhead, R. Whitham: Analyst, 1975 100, 555
21. D.R. Thomerson, W.J. Price: Analyst, 1971, 96, 1149, 825
22. G.G. Welcher, O.H. Kriege: At. Abs. Newsletter, 1970, 9, 3, 61
23. S. Greenfield, I.L. Jones, C.T. Berry: Analyst 1964, 89, 713.
24. R.H. Wendt, V.A. Fassel: Anal. Chem. 1965, 37, 920
25. G.R. Harrison:"Massachusetts Institute of Technology Wavelength Tables" 1 edn. 1969, M.I.T. Press
26. H.A. Strobel: "Chemical Instrumentation", 2 edn.; 1977, USA, Addison - Wesley.

2260.3—2251.1 A.

| Wave-length | Ele-ment | Intensities Arc | Spk.,[Dis.] | R | Wave-length | Ele-ment | Intensities Arc | Spk.,[Dis.] | R |
|---|---|---|---|---|---|---|---|---|---|
| 2256.76 | La II | 2 h | 30 | – | 2253.87 | Mg II | 8 | – | Fl |
| 2256.752 | Ir I | 2 | – | .– | 2253.86 | Ni II | 100 | 300 | – |
| 2256.74 | Co | – | 35 | – | 2253.802 | Cb | 3 | – | Me |
| 2256.67 | Cr | – | 5 | – | 2253.776 | Co | 10 | – | – |
| 2256.573 | Co I | 10 | – | – | 2253.74 | Pd | – | [2] | Bx |
| 2256.56 | Xe II | – | [3] | Hu | 2253.66 | Ni II | – | 10 | – |
| 2256.51 | Ta | 6 | 18 | a | 2253.655 | Pd | 2 | 25 | – |
| 2256.433 | Fe II | – | 10 | – | 2253.64 | Ru | 50 | – | a |
| 2256.43 | Ir | – | 10 | – | 2253.59 | Zn II | – | [3] | Vs |
| 2256.26 | Hf II | 1 | 2 | Me | 2253.55 | Ni I | 10 | – | – |
| 2256.227 | Ir | 5 | – | Ab | 2253.51 | W | – | 3 | – |
| 2256.22 | Re | 40 | 7 | a | 2253.50 | Co | – | 12 | – |
| 2256.21 | W II | 6 | 4 | – | 2253.50 | Nd | – | 25 | – |
| 2256.146 | Ni II | – | 10 | – | 2253.487 | Ir | 10 | 3 | – |
| 2256.11 | Cd | – | [2] | Bl | 2253.46 | Au II | – | 5 | – |
| 2256.104 | Pt II | – | 20 | Sh | 2253.457 | Ag II | – | 50 wh | – |
| 2256.070 | Cb | 2 | – | – | 2253.375 | Ir | 4 | – | Ab |
| 2256.05 | Cr | – | 50 | – | 2253.34 | Sr I | 3 | [3] | Fl |
| 2256.018 | W | – | 10 | – | 2253.31 | Cb | 2 h | – | – |
| 2256.00 | Ge I | 2 | 1 | – | 2253.28 | Ta | 2 | 12 | – |
| 2255.980 | Fe II | 1 | 3 | – | 2253.272 | Zr | 15 | – | – |
| 2255.91 | Au II | – | 5 | – | 2253.25 | Ti II | 1 | 12 | – |
| 2255.878 | Ni I | 10 | – | – | 2253.23 | Ir | – | 25 | – |
| 2255.860 | Fe I | 20 | – | I | 2253.18 | Mo | – | 25 | – |
| 2255.847 | Os | 125 | 2 | – | 2253.16 | Cl | – | [30] | Ks |
| 2255.810 | Ir I | 25 | 10 | – | 2253.130 | Pt II | – | 25 | – |
| 2255.788 | In II | – | [10] | Ps | 2253.124 | Fe II | 12 | 30 | I |
| 2255.77 | Ta | 3 l | 20 | – | 2253.06 | Zn II | – | [10] | Vs |
| 2255.77 | Cd | – | [2] | Bl | 2253.040 | Ir I | 10 | 2 | – |
| 2255.763 | Fe II | – | 20 | – | 2253.02 | W | – | 3 | – |
| 2255.75 | Re | 30 | 7 | a | 2253.0 | K | – | [5] | Ml |
| 2255.72 | W II | 2 h | – | – | 2253.00 | Os | 10 | – | – |
| 2255.694 | Rh | – | 15 | – | 2252.950 | V | – | 8 | – |
| 2255.65 | Co | – | 4 | – | 2252.90 | O II | – | [10] | Fl |

Table 1   Extract from MIT wavelength tables (25)

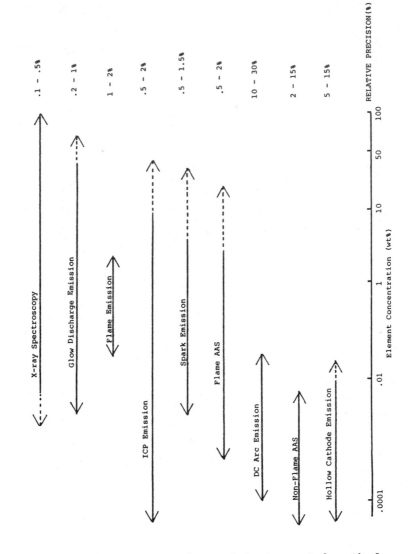

Table 2   Overall comparison of instrumental methods –
concentration range and precision

| ELEMENT | TYPICAL LOWER REPORTING LIMIT wt % |
|---------|-----------------------------------|
| Al | $.002$ |
| Co | $.005^{(1)} - .015^{(2)}$ |
| Cr | $.0015$ |
| Cu | $.0001$ |
| Mn | $.002$ |
| Ni | $.006$ |
| V | $.003 - .010$ |

Table 3  Typical lower reporting limits for trace elements in steels by flame emission spectroscopy

(1) Measured with background correction

(2) No background correction employed

| ELEMENT | METHOD A LOWER LIMIT ppm wt | METHOD B LOWER LIMIT ppm wt |
|---------|----------------------------|------------------------------|
| Ag | 0.1 | 0.002 |
| As | 50 | 1 |
| Au | 5 | int. std. |
| Bi | 0.5 | 0.01 |
| Cd | 1 | 0.2 |
| Cu | int. std. | 0.005 |
| Ge | 5 | 0.02 |
| In | 1 | 0.02 |
| Mo | 1 | - |
| Pb | .5 | 0.2 |
| Sb | 5 | 0.5 |
| Sn | 0.5 | 0.1 |
| Te | - | 1 |
| Tl | 1 | 0.05 |
| RSD | 5 - 10% | 10 - 25% |

Method A - Separation and preconcentration by coprecipitation with Cu as mixed sulphides (4).

Method B - Separation and preconcentration by coprecipitation with Mo as mixed sulphides (5) NB Best lower reporting limit reported (i.e. triple exposure plus most sensitive wavelength)

Table 4   Lower concentration limits for separation - preconcentration method and dc arc point-to-point emission spectroscopy

| ELEMENT | WAVELENGTH/Å | ELEMENT | WAVELENGTH/Å |
|---------|--------------|---------|--------------|
| P | 1782 x 2 | Mg | 2852 x 2 |
| B | 1826 x 2 | Cr | 2862 x 2 |
| C | 1931 x 2 | Mo | 2909 |
| Se | 1961 x 2 | Mn | 2933 |
| W | 2099 x 2 | He | 2945 x 2 |
| Te | 2142 x 2 | Cu | 2961 x 2 |
| Ni | 2185 x 2 | Cu | 2989 |
| Cu | 2242 x 2 | Bi | 3067 |
| Co | 2286 x 2 | V | 3110 |
| Ni | 2316 x 2 | Sn | 3175 |
| As | 2350 x 2 | Nb | 3195 |
| Ni | 2437 x 2 | Cd | 3261 |
| Si | 2516 x 2 | Ag | 3280 |
| Al | 2568 x 2 | Zn | 3345 |
| Co | 2580 | Ti | 3372 |
| Sb | 2598 x 2 | Zr | 3438 |
| Ta | 2675 | Tl | 3519 |
| Fe | 2714 | Fe | 3719 |
| Fe | 2730 x 2 | Al | 3961 |
| Hf | 2773 x 2 | Sn | 4172 |
| Mo | 2816 | Ca | 4227 |
| Pb | 2833 | In | 4511 |

X 2 indicates that a wavelength is measured in the 2nd order diffraction
Grating:  Ruled concave reflecting with 1440 lines/mm
Reciprical Linear Dispersion: 6.95 Å/mm in 1st order

Table 5  Wavelength programme for twin stand (spark
and hollow cathode) ARL B34000 optical emission
spectrometer at Glossop Superalloys Ltd

| ELEMENT | X-RAY SPECTROSCOPY Wt % | SPARK EMISSION SPECTROSCOPY Wt % |
|---------|-------------------------|----------------------------------|
| C | - | .005 |
| Si | .01 | .005 |
| Mn | .02 | .005 |
| P | .005 | .003 |
| B | - | .0005 |
| Al | .01 | .002 |
| Co | .01 | .002 |
| Cr | .01 | .005 |
| Cu | .02 | .02 |
| Fe | .02 | .03 |
| Hf | .02 | .05 |
| Mo | .01 | .005 |
| Nb | .01 | .005 |
| Ni | .01 | .005 |
| Ta | .02 | .01 |
| Ti | .01 | .002 |
| V | .01 | .01 |
| W | .01 | .05 |
| Zr | .005 | .005 |

Table 6   Comparison of lower reporting limits for
X-ray and spark emission spectroscopy

X-Ray Spectroscopy  - data obtained from Philips PW1400 sequential
                      X-Ray spectrometer

Spark Emission      - data obtained from ARL B3400 simultaneous

Spectroscopy          twin stand optical emission spectrometer

| ELEMENT | B34000 OES ANALYSIS | | | PW1400 X-RAY ANALYSIS | | |
|---------|------|------|------|------|------|------|
| | MEAN $\bar{x}$ Wt % | PRECISION 1s Wt % | RELATIVE PRECISION % | MEAN $\bar{x}$ Wt % | PRECISION 1s Wt % | RELATIVE PRECISION % |
| C | .154 | .0048 | 3.14 | (.156)[1] | (.0012) | (0.8) |
| Al | 5.48 | .053 | .96 | 5.42 | .020 | .37 |
| B | .015 | .0004 | 2.99 | - | - | - |
| Co | 9.68 | .16 | 1.68 | 9.77 | .018 | .19 |
| Cr | 9.02 | .05 | .58 | 8.84 | .007 | .08 |
| Fe | .122 | .0017 | 1.36 | .127 | .004 | 3.18 |
| Hf | 1.30 | .014 | 1.06 | 1.41 | .011 | .77 |
| Mo | .09 | .0018 | 1.99 | .114 | .0014 | 1.21 |
| Ni | 60.85 | .249 | .41 | 59.91 | .031 | .052 |
| Ta | 2.54 | .03 | 1.17 | 2.64 | .013 | .48 |
| Ti | 1.58 | .01 | .66 | 1.55 | .007 | .45 |
| W | 10.1 | .068 | .67 | 10.07 | .028 | .27 |
| Zr | .050 | .0005 | 1.12 | .042 | .0011 | 2.52 |

Statistics calculated from results of 10 consecutive analysis performed on sample of a nickel-base casting alloy.

(1) Analysis performed on a Leco CS244L Carbon and Sulphur Analyser

Table 7  Short-term precision of spark emission spectroscopy versus X-ray and combustion techniques

| ELEMENT | B34000 OES ANALYSIS | | | PW1400 X-RAY ANALYSIS | | |
|---------|------|------|------|------|------|------|
| | MEAN $\bar{x}$ wt % | PRECISION 1s wt % | RELATIVE PRECISION % | MEAN $\bar{x}$ wt % | PRECISION 1s wt % | RELATIVE PRECISION % |
| C | .158 | .010 | 6.20 | - | - | - |
| Al | 5.47 | .037 | .68 | 5.49 | .032 | .58 |
| B | .015 | .0005 | 3.5 | - | - | - |
| Co | 9.71 | .22 | 2.31 | 9.81 | .025 | .25 |
| Cr | 8.95 | .138 | 1.54 | 8.82 | .018 | .20 |
| Fe | .125 | .013 | 10.40 | .125 | .0045 | 3.58 |
| Hf | 1.35 | .031 | 2.30 | 1.46 | .018 | 1.23 |
| Mo | .09 | .01 | 11.11 | .110 | .005 | 4.55 |
| Ni | 60.02 | .31 | .52 | 59.87 | .056 | .094 |
| Ta | 2.59 | .062 | 2.4 | 2.68 | .018 | .68 |
| Ti | 1.57 | .05 | 3.18 | 1.53 | .0092 | .61 |
| W | 10.01 | .22 | 2.20 | 10.04 | .028 | .28 |
| Zr | .053 | .002 | 3.78 | .042 | .0015 | 3.6 |

Statistics calculated from results obtained by analysing sample of nickel casting alloy once per week for 8 weeks.

Table 8  Long-term precision of spark emission versus X-ray spectroscopy

| ELEMENT | MEAN $\bar{x}$ wt % | PRECISION 1s wt % | RELATIVE PRECISION % |
|---|---|---|---|
| C | .133 | .004 | 3.1 |
| Si | 2.55 | .053 | 2.1 |
| Mn | .36 | .003 | .8 |
| P | .034 | .0003 | .9 |
| Al | .052 | .002 | 4.3 |
| Co | .147 | .0013 | .9 |
| Cr | 21.54 | .097 | .5 |
| Cu | .46 | .005 | 1.2 |
| Fe | 61.26 | .129 | .21 |
| Mo | .44 | .003 | .7 |
| Nb | .48 | .006 | 1.2 |
| Ni | 8.69 | .069 | .8 |
| Ta | .26 | .001 | .5 |
| Ti | .009 | .0002 | 2.8 |
| W | 3.57 | .035 | 1.0 |

Table 9  Analytical precision (short term) of spark emission spectroscopy on highly-alloyed steel sample

| ELEMENT | WAVELENGTH Å | APPLICABLE CONCENTRATION RANGE   ppm (wt) |
|---|---|---|
| Ag | 3280 | 0.1 - 10 |
| As | 2349 x 2 | 5 - 100 |
| Bi | 3067 | 0.1 - 10 |
| Cd | 3261 | 0.1 - 5 |
| Ga | 4172 | 10 - 100 |
| In | 4511 | 0.1 - 5 |
| Mg | 2852 x 2 | 5 - 200 |
| Pb | 2833 | 0.3 - 30 |
| Sb | 2598 x 2 | 1 - 50 |
| Se | 1961 x 2 | 1 - 20 |
| Sn | 3175 | 5 - 100 |
| Te | 2142 x 2 | 0.5 - 20 |
| Tl | 3519 | 0.1 - 5 |
| Zn | 3345 | 2 - 20 |
| He | 2945 x 2 | Internal Standard |

Table 10  Hollow cathode emission spectroscopy;
Elements determined, wavelengths and concentration
range for ARL B34000;  OES at Glossop Superalloys Ltd

| ELEMENT | LOW-ALLOY STEEL | | | HIGH-ALLOY STEEL | | |
|---------|------|------|------|------|------|------|
| | MEAN $\bar{x}$ Wt % | PRECISION 1s Wt % | RELATIVE PRECISION % | MEAN $\bar{x}$ Wt % | PRECISION 1s Wt % | RELATIVE PRECISION % |
| C | .59 | .0098 | 1.7 | .059 | .0021 | 3.6 |
| Si | .27 | .0019 | .7 | .56 | .0041 | .7 |
| Mn | .92 | .0081 | .9 | 1.40 | .0027 | .2 |
| P | .034 | .0011 | 3.3 | .035 | .0008 | 2.3 |
| Al | .067 | .0043 | 6.4 | .008 | .0006 | 6.9 |
| As | .035 | .0006 | 1.8 | .073 | .0027 | 3.7 |
| B | .0034 | .0001 | 2.1 | .0036 | .0001 | 2.6 |
| Co | .102 | .0024 | 2.3 | .43 | .0034 | .8 |
| Cr | .45 | .0055 | 1.2 | 18.25 | .0944 | .5 |
| Cu | .21 | .0020 | .9 | .22 | .0013 | .6 |
| Fe | 96.48 | .0215 | .02 | 64.85 | .0931 | .1 |
| Mg | .0030 | 0 | – | .0041 | .0002 | 5.9 |
| Mo | .22 | .0028 | 1.3 | 2.54 | .0073 | .3 |
| Ni | .40 | .0032 | .8 | 11.41 | .0478 | .4 |
| Pb | .0054 | .0003 | 5.2 | .010 | .0002 | 1.9 |
| Ti | .098 | .0016 | 1.7 | .0011 | .0004 | 31.8 |
| Sn | .025 | .0008 | 3.3 | -.0076 | .0006 | 8.3 |
| W | .056 | .0024 | 4.3 | .11 | .0017 | 1.5 |

Table 11  Analytical performance of glow discharge emission spectroscopy (Courtesy of Hilger Analytical)

| ELEMENT | ICP | FLAME AAS | NON-FLAME AAS |
|---------|-----|-----------|---------------|
| Ag | .005 | .002 | .00005 |
| Al | .02 | .03 | .00035 |
| As | .02 | .3 | .0004 |
| B | .004 | .5 | .05 |
| Bi | .05 | .05 | .00035 |
| Cd | .002 | .0015 | .00015 |
| Co | .006 | .005 | .00025 |
| Cr | .005 | .006 | .00013 |
| Cu | .002 | .003 | .00013 |
| Fe | .003 | .006 | .0001 |
| Ga | .05 | .08 | .001 |
| Hf | .015 | 2.0 | - |
| In | .05 | .04 | .0013 |
| Mg | .0005 | .0003 | .00001 |
| Mn | .001 | .002 | .000035 |
| Mo | .005 | .02 | .0004 |
| Nb | .02 | 2.0 | - |
| Ni | .01 | .01 | .0005 |
| P | .05 | 4.0 | .1 |
| Pb | .05 | .01 | .00015 |
| Sb | .04 | .04 | .00045 |
| Se | .05 | .5 | .0013 |
| Si | .009 | .25 | .0013 |
| Sn | .03 | .1 | .0011 |
| Ta | .02 | 2.0 | - |
| Te | .05 | .03 | .001 |
| Ti | .001 | .08 | .0023 |
| Tl | .05 | .02 | .0013 |
| V | .005 | .07 | .0014 |
| W | .02 | 1.0 | - |
| Zn | .001 | .0008 | .00001 |
| Zr | .004 | 1.0 | - |

All values are in $\mu g\ ml^{-1}$

Table 12   Sensitivity of ICP, flame and non-flame AAS
(Courtesy of Philips Scientific)

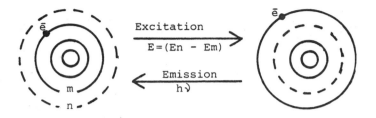

Ground state atom with
electron at Em (En not
occupied)

Excited atom with
electron at En
(Em not occupied)

Figure 1   Principle of excitation and emission

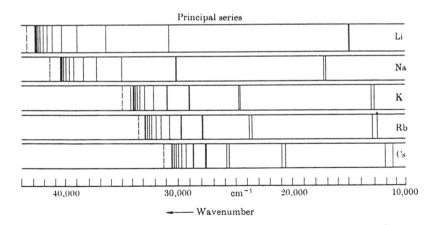

Figure 2   Principal emission series for the common
alkali metals (Courtesy of Addison Wesley Publ. Co.
Inc.)

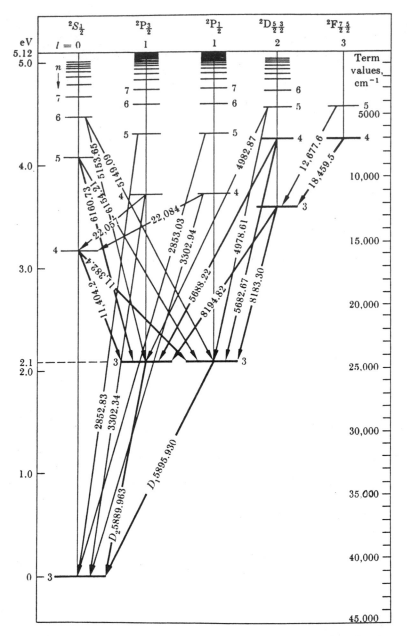

Figure 3  Term diagram for the sodium atom.  Emission
transitions are shown by arrows with their wavelengths
(Courtesy of Addison Wesley Publ. Co. Inc.)

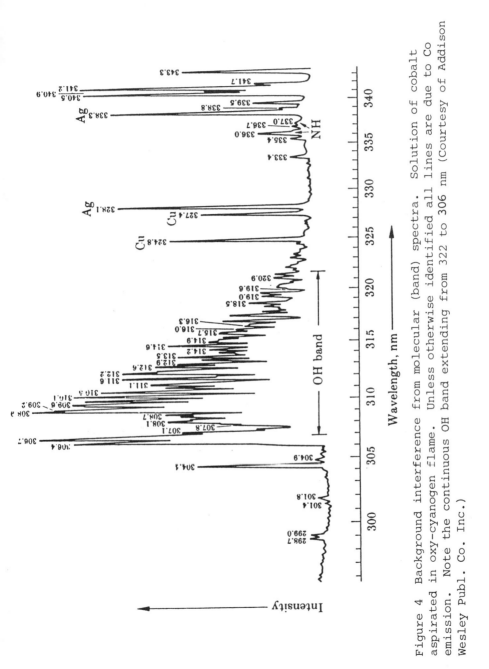

Figure 4 Background interference from molecular (band) spectra. Solution of cobalt aspirated in oxy-cyanogen flame. Unless otherwise identified all lines are due to Co emission. Note the continuous OH band extending from 322 to 306 nm (Courtesy of Addison Wesley Publ. Co. Inc.)

53

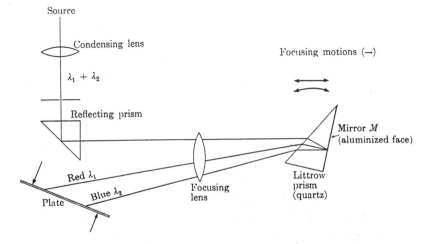

Figure 5  Littrow quartz prism emission spectrograph with photographic detection (Courtesy of Addison Wesley Publ. Co. Inc.)

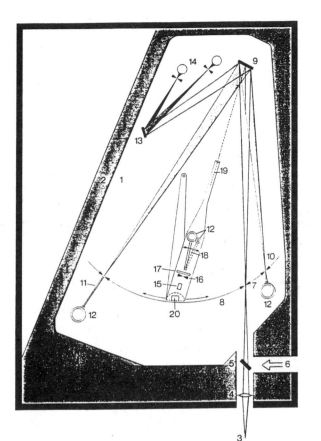

1. Constant low-pressure argon zone
2. Temperature-controlled zone
3. Analytical gap of excitation stand
4. Entrance lens
5. Switchable mirror(cylindrical for ICP system)
6. Light from ICP mirror optics or second excitation position
7. Entrance slit(motorised displacement with automatic background correction option.
8. Exit slit plate
9. Concave primary grating
10 Exit slit on long wavelength side of entrance slit
11. Light guide
12. Photomultiplier
13. Plane secondary grating
14. Long wavelength exit slits and photomultipliers.
15. Angled plane mirror
16. Exit slit
17. Filter turret
18. Switchable beam
19. Tie bar
20. Position reading head.

Figure 6   Modern polychromator based on Rowland circle concept.  Light guides are used to convey dispersed wavelengths to the photomultiplier tubes.  The instrument also features twin stand **configuration**, split optical arrangement and scanning channel. (Courtesy of Philips Scientific)

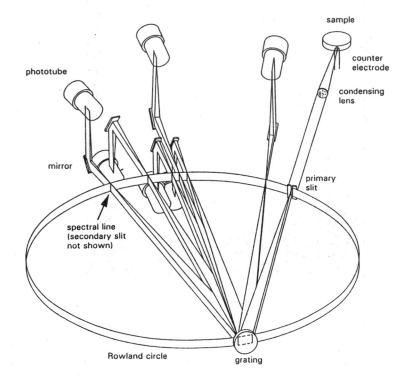

sample

counter
electrode

phototube

condensing
lens

mirror

primary
slit

spectral line
(secondary slit
not shown)

Rowland circle

grating

Figure 7  Modern polychromator based on Rowland circle
concept;  Torroidal mirrors are used to focus
individual wavelengths onto photomultiplier tubes
(Courtesy of Applied Research Laboratories Ltd)

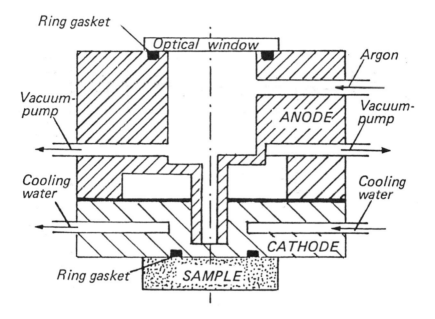

Figure 8   Glow discharge lamp (Courtesy of Applied
Research Laboratories Ltd)

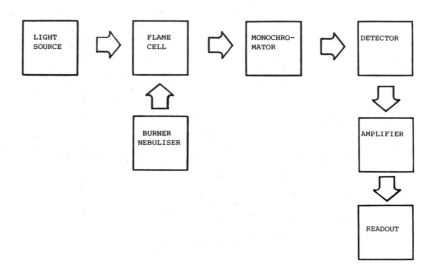

Figure 9 Block diagram of a flame AAS system

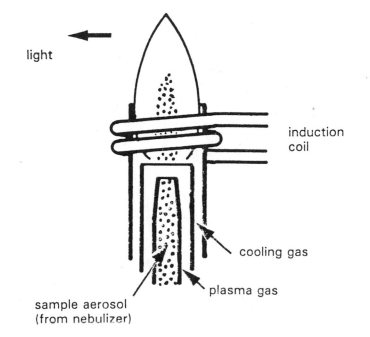

light

induction
coil

cooling gas

plasma gas

sample aerosol
(from nebulizer)

Figure 10    Inductively coupled plasma torch (Courtesy
of Applied Research Laboratories Ltd)

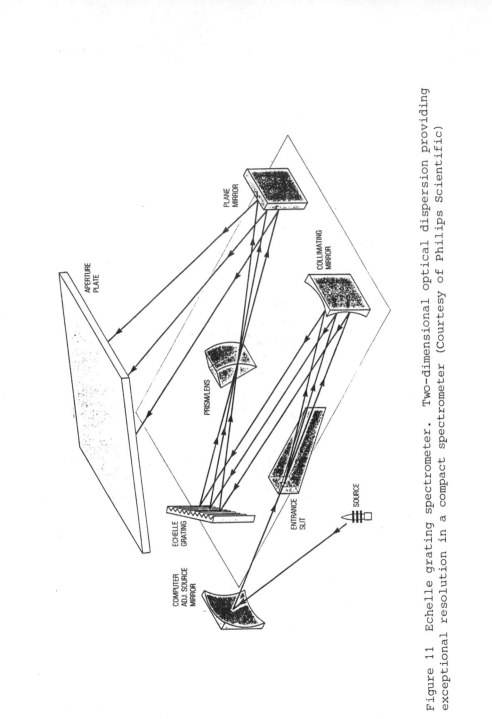

Figure 11 Echelle grating spectrometer. Two-dimensional optical dispersion providing exceptional resolution in a compact spectrometer (Courtesy of Philips Scientific)

# 3 : Determination of gases in high-temperature materials
# K THORNTON

The author is with Inco Alloys Ltd, Hereford

## PART I
### INTRODUCTION

Producers and users of high-temperature materials are
mainly interested in the levels of oxygen and nitrogen,
although for certain alloys the level of hydrogen is
important. In addition there are some specialized
production procedures during which argon may be
introduced and this gas then becomes important. In this
presentation techniques will be discussed for the
determination of each of these four gases.

## 1. DETERMINATION OF HYDROGEN

Hydrogen is introduced into high-temperature materials
at the melting stage, mainly through humidity of the
atmosphere and refractory materials and also from some
of the raw materials, particularly those produced by
electrolytic procedures, such as manganese. It is
present interstitially and in voids, but can also be
present in the form of hydrides, particularly in
titanium and zirconium alloys. The level is generally
between 0.2 ppm and 10 ppm in most steels and nickel
base alloys, but can be significantly higher in titanium

alloys - typically 20 ppm to 150 ppm.

In many materials where it is present interstitially it has high mobility and will diffuse from the sample by heating to temperatures considerably below the melting point. For instance, hydrogen is mobile in ferritic steels at temperatures as low as -80°C, and there is some limited mobility at room temperature for austenitic steels. This feature leads to difficulties in the manufacture and storage of stable reference materials.

There are two main techniques for the determination of hydrogen and these are described below.

## A.   Hot Vacuum Extraction

In this procedure the high mobility of hydrogen is used by heating the sample in a vacuum at a temperature below the melting point of the alloy. Temperature is selected to promote adequate mobility so that the hydrogen will diffuse from the sample in a reasonable time, but excessive temperatures are avoided in order to minimize the liberation of other gases from the sample. The equipment is illustrated in Figure 1.

The sample is introduced into a loading port and the apparatus is evacuated by means of a rotary pump. The sample is then dropped into a quartz tube mounted in a resistively heated furnace and maintained at the optimum temperature for diffusion of hydrogen from the alloy. Typical analysis temperatures are given in Table 1. After a pre-determined time the evolved gases are collected in a calibrated volume and the pressure is measured. The gases are then transferred to the inside of a heated palladium thimble where hydrogen rapidly diffuses through to the outside. The remaining gases are returned to the calibrated volume and the pressure is measured again. The difference between the two pressure readings is a direct measure of the evolved hydrogen.

This procedure is an absolute one, relying only on physical constants and calibrated volumes. It is therefore authoritative and used as an independent referee procedure.

Although the technique is elegant, there are a number of problems. The equipment is mainly made of glass and can be fragile, extraction times are lengthy (typically 5 to 30 minutes) and the need for a good vacuum is important. In addition some problems arise if high levels of other diffused gases are present.

## B.   Carrier Gas Fusion

Because of the time factor for hot vacuum extraction, emphasis has recently moved to the determination of hydrogen by techniques involving fusion of the sample. Both vacuum fusion and carrier gas fusion have been used but the latter is more common.

In this technique a sample weighing between 0.1g and 10g, is heated in a graphite crucible in a stream of 'carrier gas'. The temperature is usually maintained between $1600°C$ and $2000°C$, depending on the melting point of the alloy. Under these conditions hydrogen is rapidly evolved together with nitrogen and carbon monoxide. The gases are swept out of the furnace region in the carrier gas stream and passed through a series of reagent columns to remove certain impurities. Carbon monoxide is removed by oxidation to carbon dioxide with Schutze reagent (iodine pentoxide) and subsequently absorbed in soda-asbestos. Water vapour is absorbed by Anhydrone (anhydrous magnesium perchlorate).

If nitrogen is used as the carrier gas, the gas stream then consists of nitrogen and hydrogen. This passes directly to the detector.

Alternatively, where argon is used as the carrier gas, the gas stream at this stage consists of argon, nitrogen and hydrogen. In this case the nitrogen and hydrogen must be separated by the passage of the gas stream through a column of molecular sieve. This retards the impurities in proportion to their molecular weight. Hence the argon carrier gas will contain discrete 'packets' of hydrogen and nitrogen.

Hydrogen is measured using a thermal conductivity detector, which consists of four heated wire filaments connected in a Wheatstone Bridge network. Two of the filaments are situated in the pure carrier gas and the other two are in the analysis gas stream. When no impurities are present the system is balanced. When an impurity is introduced with a thermal conductivity differing from that of the carrier gas, the temperature of the filament wires in the analysis arms of the bridge will change. This causes a change in resistance which can be measured as a d.c. voltage and, with suitable treatment, can be converted to display the hydrogen

63

concentration. A schematic representation using argon as the carrier gas is given in Figure 2.

For maximum sensitivity, the carrier gas should differ in thermal conductivity from hydrogen by as much as possible. Thermal conductivities for some gases are given in Table 2, from which it can be seen that the best sensitivity is obtained using argon as the carrier gas. Calibration of the equipment is performed by gas-dosing, i.e. the introduction of known volumes of hydrogen.

Compared with the hot vacuum extraction procedure this method is faster - approximately 1 to 3 minutes per sample - and is therefore finding increased popularity. However, it should be recognised that the two techniques do not necessarily give the same results. In the first procedure only molecular and interstitial hydrogen is measured. However, in the fusion procedure some hydrogen compounds may be dissociated in the high-temperature reducing atmosphere and contribute to the measurement.

When analysing ferritic and low-alloy steels it is necessary to guard against loss of hydrogen from the samples prior to analysis. For these materials the rate of diffusion of hydrogen is significant at room temperature and reliable results can only be obtained if the specimens are transported and stored in liquid nitrogen.

## 2.   DETERMINATION OF OXYGEN

Oxygen is almost invariably determined by fusion techniques, either under vacuum or in the presence of a carrier gas. Oxygen is converted to carbon monoxide by reduction of the oxides in the sample with carbon.

$$MO + 2C \rightleftharpoons MC + CO$$

This reaction is driven to the right by a large excess of carbon from the graphite crucible in which the sample is fused and by reduction of the partial pressure of carbon monoxide above the melt. This is achieved by vacuum pumping or by transport of the carbon monoxide away from the reaction zone in a carrier gas.

For carrier gas systems a schematic representation of the furnace is given in Figure 3. The graphite crucible is pre-burned to remove atmospheric and other

contaminants, after which a sample, weighing typically
0.5 to 1.0g, is introduced and melted in a stream of
helium carrier gas.   Under these conditions carbon
monoxide is liberated according to the above equation,
together with nitrogen and hydrogen.  Measurement of the
carbon monoxide, and hence the oxygen content of the
sample, may be done in a number of ways as shown below:-

(a)  Direct measurement of CO by infra-red absorption

(b)  Oxidation of CO to $CO_2$, followed by infra-red
     absorption

(c)  Oxidation of  CO  to  $CO_2$,  followed  by  thermal
     conductivity

The first procedure is simple, but may be prone to error
if very high oxygen contents are encountered, because
some of the oxygen may be converted into carbon dioxide
in the furnace and would not be measured.   The second
two procedures avoid this problem by ensuring that all
oxygen is converted to carbon dioxide downstream of the
furnace.
     The third procedure is often used simply because it
offers the possibility of a single detector for both
nitrogen  and  oxygen  in  a  combined  gas  analyser.
However, if this option is chosen it is necessary to
separate the carbon dioxide from other gases by passage
through  a  column  of  molecular  sieve.   A  schematic
representation of one manufacturer's combined oxygen and
nitrogen system is given in Figure 4.   In this case the
oxygen  is  measured  by  infra-red  absorption  after
conversion to carbon dioxide.
     Liberation  of  oxygen  from  metal  oxides  is  an
endothermic reaction and requires high temperatures for
dissociation and formation of carbon monoxide. The most
refractory oxides, such as silica and alumina require
heating  to  temperatures  in  excess  of  2000°C  for  the
reaction to proceed.  For this reason, impulse furnaces
are preferred to induction furnaces.  The former operate
at low a.c. voltages ( $\sim$ 10 volts) and high currents (up
to 1200 amps) and are capable of heating the sample to
temperatures of about 3000°C.   Induction furnaces are
generally limited to about 2000°C.
     Calibration of carrier gas fusion instruments is
usually achieved by gas - dosing with carbon monoxide or

carbon dioxide, whichever is appropriate to the measuring system. This is satisfactory for calibration of the detector, but does not take account of the extraction efficiency of the furnace system. For this reason, the basic calibration is checked by analysing certified reference materials. Unlike hydrogen, oxygen is stable in high-temperature material reference samples, assuming that adequate precautions are taken to protect or remove the outer surface.

Difficulties are sometimes experienced in the control of blank levels. These are critically dependent on maintaining a good gas-tight system and ensuring thorough cleanliness of the total system, including gas purification reagents. With modern equipment it is now possible to maintain blank levels below the equivalent of 1 ppm oxygen.

However, it can be very difficult and time consuming to obtain a representative sample from the material to be analysed, largely because of spurious oxidation problems during sample cutting. Various chemical cleaning and etching procedures have been tried, but these can be lengthy and are not comprehensively useful for the range of materials encountered. Direct analysis of millings or drillings, even when prepared at low cutting speeds will lead to erroneously high results because of surface oxidation. The most successful preparation procedures involve cutting a solid specimen of about 1g and cold filing this by hand to remove all of the pre-existing surface. The sample is finally washed in acetone. Using this procedure and extraction by carrier gas fusion, a recent inter-laboratory exercise, co-ordinated by Laboratory of the Government Chemist, has produced repeatability values of about ±2 ppm for oxygen. This suggests that, with the present technology, the limit of detection for oxygen is in the order of 5 parts per million.

Alternative methods are available for oxygen, but these find little use in the laboratory. In particular, neutron activation analysis using 14 MeV neutrons has been used, but the limit of detection is relatively poor (approx 10-20 ppm oxygen). Within the melting department, solid 'probe' analysers are sometimes used for determination of oxygen in the molten metal. These are based on $ZrO_2/CaO$ solid electrolytes which produce a voltage proportional to the activity of oxygen in the melt.

## 3.   DETERMINATION OF A NITROGEN

There are two principal techniques for the determination of nitrogen in high-temperature materials, the one based on reductive fusion and the other on chemical methods after conversion to ammonia.

## A.   Reductive Fusion

Methods in routine use for the determination of nitrogen in high-temperature materials are essentially similar to the fusion techniques used for oxygen determination. Indeed, many instruments using vacuum fusion or carrier gas fusion are designed for the simultaneous determination of both gases. However, there are some important considerations that apply specifically to nitrogen:-

Nitrogen is not easily measured by infra-red absorption, so that thermal conductivity is invariably used. This necessitates the removal of other impurity gases from the carrier gas stream prior to measurement. To accomplish this, carbon monoxide is oxidized to carbon dioxide and subsequently absorbed on soda-asbestos. Hydrogen is converted to water vapour over a heated copper oxide catalyst and absorbed in Anhydrone. The carrier gas then contains only nitrogen and is passed to the thermal conductivity detector. From Table 2 it is clear that the best sensitivity is obtained by using helium as the carrier gas (hydrogen would present too many safety problems).
    Secondly, although sensible precautions must be taken during sample preparation, it is not necessary to be as rigorous as for the preparation of oxygen samples. Satisfactory results can be obtained from specimens prepared as millings or drillings provided that overheating is avoided.
    Thirdly, in most high-temperature materials nitrogen is present as discrete nitride or carbonitride phases. These are often very stable compounds, particularly when present as nitrides of titanium, boron, silicon and aluminium. The presence of these stable compounds requires high temperatures in the graphite crucible for quantitative recovery. In some cases, for instance titanium nitride, it is still not possible to obtain complete dissociation unless

significant amounts of 'diluent' elements are present. Fortunately, in most high-temperature alloys the major elements provide sufficient dilution of the titanium to avoid this problem.

Finally, nitrogen is slow to diffuse from the melt and this makes it difficult to drive the dissociation reaction to completion. Furthermore, as the melt temperature increases graphite is taken into solution and precipitates within the molten sample. This increases the viscosity and futher reduces the rate of diffusion. To counteract this, additions may be made to the crucible to promote formation of spheroidal graphite rather than the flake form. The former has less effect on viscosity and leads to improved nitrogen recovery.

From this discussion it is seen that, if problems arise, they are generally likely to lead to erroneously low results for nitrogen. This is the reverse for oxygen determination, where the main problems are sample preparation contamination, which can lead to erroneously high oxygen results.

However, with the exception of titanium alloys, fusion methods are now capable of providing accurate and reliable values for nitrogen in high-temperature materials. Precision is about ±0.5 ppm at 10 ppm nitrogen and about ± 3 ppm at 100 ppm nitrogen, with a limit of detection of 1-2 ppm.

## B.   Kjeldahl Chemical Procedure

For nitrogen we are fortunate to have an independent chemical method available, based on a procedure developed by Kjeldahl. In this technique the sample is dissolved with hydrochloric acid whereupon nitrogen is converted to the ammonium ion, $NH_4^+$. Dissolution is carried out under an air condenser to prevent loss of $NH_4^+$ to atmosphere. Any insoluble residue is heated strongly with sulphuric acid to break down remaining nitrides. The ammonium ion is then steam distilled after addition of sodium hydroxide, collected in a receiving solution of boric acid and titrated against standard hydrochloric acid solution. The procedure is shown schematically in Figure 5.

Although used successfully for titanium alloys and for the independent analysis of a number of steel certified reference materials, the method is not widely used for determination of nitrogen in high-temperature

68

materials. This is largely because of the time factor, but also because of some technical problems. The presence of stable nitrides makes it necessary to use vigorous dissolution techniques to recover all of the nitrogen. Under these conditions, variability increases, together with blank values. For simple materials where stable nitrides are absent the precision is about ± 3 ppm at 30 ppm nitrogen, but for complex alloys the precision is unlikely to be better than ± 10 ppm at this level.

## 4.   DETERMINATION OF ARGON

The requirement for determining argon is limited to those materials where the production cycle is likely to introduce argon into the metal. These include argon powder atomization and hot isostatic pressing (HIP) where argon is used as the pressurized fluid. Argon is totally inert and is therefore present interstitially or in voids. The inertness of the gas makes its determination relatively simple.

As for oxygen and nitrogen, methods are based on carrier gas fusion, but the temperature for complete liberation is generally lower than in the previous two cases. The sample is melted in a graphite crucible at a temperature of about 2000°C and the argon is carried away from the furnace in a stream of carrier gas. Detection is by thermal conductivity which requires that other impurities are removed from the gas stream. As before, carbon monoxide is converted to carbon dioxide and absorbed on soda-asbestos, and hydrogen is converted to water vapour and absorbed on Anhydrone. Nitrogen and argon remain in the carrier gas and these are time-resolved by passage through a column of molecular sieve.

Reference to Table 2 shows that helium is the best carrier gas for maximum sensitivity. Using this system it is possible to determine argon down to about 0.2 ppm using a 1g sample.

## 5.   GAS SPECIATION

Oxygen and nitrogen are present in solid solution and as discrete compounds in many high-temperature materials. In steels the former may be present as $FeO$, $Al_2O_3$, $SiO_2$ and nitrogen may be present in simple or complex

69

nitrides. The decomposition of these species is governed by thermodynamic equilibria such that each has a characteristic minimum temperature under given operating conditions. This leads to the possibility of separately identifying such compounds by gradually increasing the furnace temperature in a controlled programme.

Equipment is now available to perform this type of analysis and there is considerable interest in the totally new information that these results may provide. In a typical programme the sample is introduced into a graphite crucible which is heated through the range 1200°C to 2600°C in a carrier gas atmosphere. The rate of temperature increase is adjusted for optimum separation of peak signals, but is generally 5 to 10°C per second. A continuous display of the oxygen or nitrogen signal is plotted against time or temperature.

An example of this type of analysis is shown in Figure 6, which is a record of the evolution of oxygen from a nickel-base alloy produced by mechanical alloying. Two dissociation temperatures are clearly observed suggesting the presence of at least two different oxide species.

This important development is still in its infancy and there are still considerable problems awaiting resolution before individual oxides or nitrides can be unambiguously identified and quantified.

## CONCLUSIONS

Techniques for the determination of hydrogen, oxygen, nitrogen and argon are readily available, mostly based on fusion - extraction principles. The predominant techniques employ carrier gas fusion in a graphite crucible, rather than vacuum fusion. These procedures are generally reliable, except for the determination of nitrogen in titanium alloys, where the Kjeldahl chemical method should be used.

We are now entering a challenging period during which the analyst may learn to provide information on specific nitrides and oxides rather than just total oxygen or nitrogen.

## Table 1

## Vacuum extraction temperatures for determination of hydrogen

| MATERIAL | EXTRACTION TEMPERATURES, °C |
|----------|------------------------------|
| Steel, low alloy | 600 |
| Steel, austenitic | 1000 |
| Nickel alloy | 1050 |
| Titanium | 1120 |

## Table 2

## Thermal conductivities for some gases

| MATERIAL | THERMAL CONDUCTIVITY* |
|----------|------------------------|
| Hydrogen | 39.6 |
| Helium | 33.6 |
| Nitrogen | 5.6 |
| Argon | 3.8 |
| Carbon dioxide | 3.4 |

*Cal/cm/sec x $10^5$

## Figure 1
## Hot Vacuum Extraction – Determination of Hydrogen

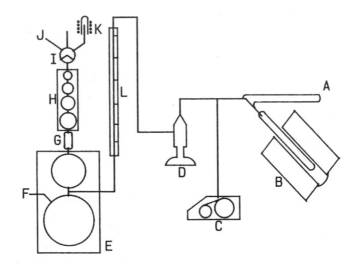

A-Sample tube    B-Furnace    C-Rotary vacuum pump
D-Diffusion pump E-Toepler pump   F-Air release
G-Float valve    H-Calibrated volumes    I-Stopcock
J-To vacuum pump    K-Heated Palladium Thimble
L-Mercury in glass pressure gauge

Figure 2
Carrier Gas Fusion for the determination of Hydrogen

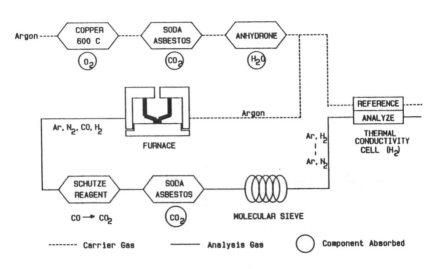

Figure 3
Impulse Furnace for Carrier Gas Fusion

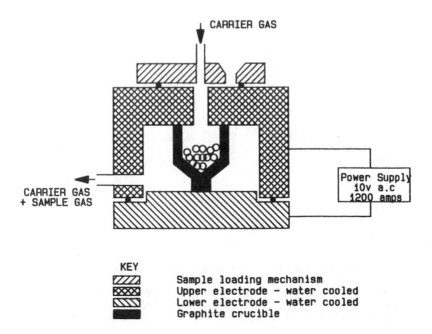

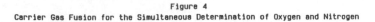

Figure 4
Carrier Gas Fusion for the Simultaneous Determination of Oxygen and Nitrogen

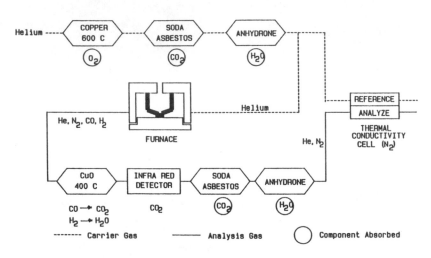

Figure 5
Kjeldahl Method for Determination of Nitrogen

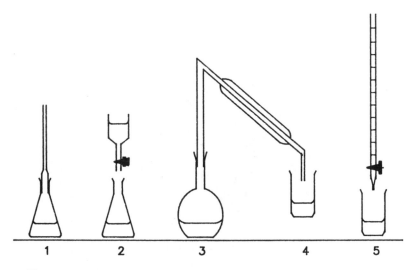

**Key**
1 Sample dissolved in acid under air—cooled condenser
2 Sodium hydroxide solution added
3 Ammonium ion steam distilled from sample solution
4 Condensate containing ammonium ion collected in boric acid
5 Ammonium ion titrated against standard hydrochloric acid

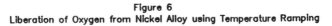

**Figure 6**
**Liberation of Oxygen from Nickel Alloy using Temperature Ramping**

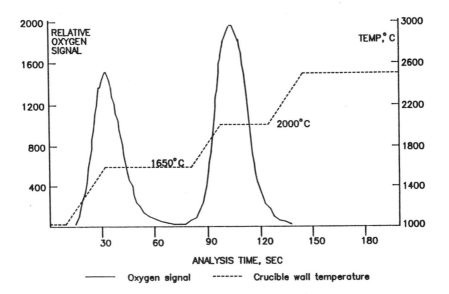

## PART II
### INTRODUCTION

Knowledge and control of trace element levels has become an increasingly important factor in the manufacture of materials for use in high stress, high temperature applications. This is reflected in the increasingly severe restrictions for a variety of elements in material specifications. Table 1 clearly illustrates the change in requirements for cast nickel superalloys since 1975 and shows a dramatic increase in the number of elements specified as well as reductions in the limits for those that were originally controlled. From the alloy user's standpoint the most important consideration is the effect of these elements on the service properties of the materials, but it is important to recognize that the alloy manufacturer is equally concerned because of the effects on metal working properties.

The concept of a 'trace' element has different interpretations in different industries; for example, in the semiconductor field we may restrict ourselves to levels in the parts per billion range, for aircraft engine superalloys the interest is generally in the part per million range, whereas for some of the less demanding engineering applications for steels the interest may extend only to tens or hundreds of parts per million. For the purpose of this review we shall concentrate on those elements that are present at up to about 100 ppm.

The inclusion of trace elements in specifications should ideally be based on theoretical predictions or empirical knowledge of their effects on material properties. However, there is little evidence that this approach is adopted in many instances. The user's specifications are often based on the assumption that impurity elements are deemed to be undesirable unless their presence is proved to be innocuous. This is an understandable position since many of the materials are used in critical applications where lack of integrity may have significant safety implications. However, this approach leads to major quality control problems in the manufacture of material where large numbers of trace elements are included in the specification, as for instance in AMS 2280.

76

Although most effort is directed towards the
control and measurement of deleterious trace elements,
it is important to recognise that some elements at
trace levels cause beneficial effects; for example the
addition of magnesium to some nickel-base superalloys is
used to improve hot working characteristics by combining
with the otherwise deleterious element, sulphur. A
review of the known effects of trace elements in steels
and nickel alloys is given by Mayer and Clark[1].

The objective of this paper is to provide a broad
over-view of the techniques that are currently available
and in use for the determination of trace elements in
high-temperature materials. Most of the methods
discussed are of wide applicability, both in terms of
the range of materials that can be analysed and the
number of elements that are determinable. It is
recognized that there are significant omissions,
particularly in the fields of neutron activation, X-ray
fluorescence and classical chemical procedures, all of
which have an important role in particular fields.

## DETECTABILITY AND ACCURACY

In trace element determinations, the most important
concept is the detection limit (DL), which is defined
as:

$$DL = 3 \times \sigma_{BGD}$$

where $\sigma_{BDG}$ is the standard deviation of ten to
twenty measurements at or near the
background level.

For good detectability a method must be
characterized by low background levels, low variability
of readings at the background level and high sensitivity
(signal per unit concentration) - see Figure 1.

Analytical precision deteriorates as the detection
limit is approached. At high concentrations the
precision improves from a limiting value of ±33%
relative in an exponential manner, as shown in Figure
2. Obviously, at or near the detection limit precise
quantitative analysis is not possible. This leads to
the concept of limit of quantitative determination (LQD)
which most laboratories use, but which has no formal
definition. It is the value below which a laboratory
will not report an absolute value and generally lies

between one and ten times the detection limit. The LQD is influenced by factors other than statistical considerations, such as availability of certified reference materials.

In comparing detection limits one must be careful to obtain information which is relevant to the materials to be analysed. Manufacturers will often quote values for simple aqueous solutions and, in general, these may be much better than the values obtained for solutions of complex alloys in strong acids. This is because both variability and absolute value of the background may increase significantly in the latter case.

The relative precision required for the determination of trace elements is usually much less demanding than for alloying concentrations of elements. For instance it may be perfectly acceptable to use a method with a relative precision of ±50% at the two parts per million level, implying a value of 2 ± 1 ppm, whereas much better relative precision will be necessary at the per cent level (typically better than ±1% of content).

Precision is only one of the factors involved in the accuracy or 'trueness' of a result. At trace concentrations, unsuspected blanks and specific interferences may cause considerable errors, particularly where certified reference materials closely matched with the samples are unavailable. Unfortunately, the supply of CRM's for trace elements in high-temperature materials is always likely to be inadequate and most laboratories rely to a great extent on in-house standards. This may lead to bias between laboratories which will need to be resolved by inter-laboratory comparisons. Ultimately it may be necessary to agree standards and the analytical methodology, even if it is suspected that the accuracy of the results may be biased.

## SAMPLE TREATMENT

Before proceeding to a review of the individual analytical techniques it is appropriate to make a few generalizations about sample treatment.

In general, the analyst prefers to carry out as little treatment of the sample as possible. This is because every handling step has the potential to introduce contamination. As we have seen, increase in the level and variability of the blank causes a deterioration in the detection limit. In those cases where acid dissolution of the sample is necessary, the ratio of acid to sample should be kept as low as possible. If alternative dissolving mixtures are available, the choice should involve consideration of the relative blank levels. As a fundamental requirement, each analytical run, together with the calibration standards, should use the same batches of acids. Care must also be exercised to ensure that the dissolving operation does not cause loss of the analyte through precipitation or distillation and that contamination from the laboratory environment is minimized.

Many of these problems can be reduced by dissolution in closed pressure vessels (such as PTFE) and heating in a microwave oven.

In those cases where chemical pre-treatment is unnecessary, it is still important to take steps to avoid contamination. For instance, the production of chippings or millings from solid metal samples may give rise to contamination from oils or cutting fluids. Even if these do not contain the analyte, the presence of such materials may influence the measurements - particularly in some of the spectroscopic techniques. Similarly, special cutting devices such as shaped tungsten carbide tips are often brazed on to steel shanks. In these cases it is possible to contaminate the chippings by contact with the brazing material.

It is often necessary to use pre-concentration methods before a particular analytical method is applied. This may be because the detection limit is not good enough when using the direct method, or to separate the analyte from known interfering elements in the sample. There are many such methods, some of which will be referred to in the following sections, but the analyst must balance the advantages against the possible disadvantages. These include higher blank values, incomplete recovery of the analyte and reduced throughput.


## TECHNIQUES

1.  <u>Optical Emission Spectroscopy (OES) using the d.c. arc source</u>

Many of the earliest examples concerning the effects of trace elements on high-temperature materials were originally investigated and established by use of d.c. arc - OES. The principles of optical emission spectroscopy are defined elsewhere in this monograph by Murphy, but a brief outline is included below.

The sample, in the form of a solid or powder is volatilized in an electrical discharge, in this case a direct current arc. The atoms of the individual elements that are present in the vapour are thermally excited within the high-temperature region of the arc column at temperatures between about 5000°C and 7000°C and subsequently decay to their ground state by emission of radiation in the ultra-violet and visible regions of the spectrum. This radiation is analysed by dispersion with a prism or diffraction grating and recorded on a photographic plate or by means of opto-electrical direct reading devices such as photomultipliers.

Until about 1970 this technique was the most commonly used procedure for investigation and quality control of trace elements in high-temperature materials, but its use and popularity have declined with the advent of alternative techniques such as atomic absorption spectroscopy.

If the sample is in the form of a suitable solid, a d.c. arc is struck between a flat face of the sample and a rod counter electrode such as graphite or high purity copper. In a typical configuration the voltage drop across a 5 mm arc gap running at 5 amperes is about 30 volts. Limits of detection depend on a number of factors including light-gathering power and resolution of the spectrometer, electrical discharge parameters, length of integration time, choice of spectral line and sample matrix. In Table 2 approximate limits of detection are quoted for a range of elements in a simple nickel-chromium heat-resisting alloy using this technique. These were obtained with a 3 metre Ebert grating spectrograph operating with a 5 ampere d.c. arc and 15 second integration. Photographic recording was used for the most sensitive spectral lines in the available wavelength range.

In many cases it is advantageous to use a graphite cup electrode to contain the sample, as shown in Figure 3. The sample, which may be in the form of chippings or powder, is loaded into the lower cup electrode and a d.c. arc is struck between this and a pointed counter-electrode. Advantages include the ability to analyse small samples and improved detection limits for many elements. Perhaps more importantly, it is possible to analyse mixed oxides by this technique, which provides the opportunity to make synthetic reference materials either by direct mixing or by evaporation and subsequent ignition of solutions of synthetic samples. Similar methods for a variety of materials are included in Methods for Emission Spectrochemical Analysis published by American Society for Testing and Materials. Table 3 gives the concentration ranges for twelve elements in molybdenum and molybdic oxide as described in method E-2 SM 8-21[2]. The method has good detection limits, is based on synthetically prepared reference standards and has adequate precision (8-15% relative).

The detection limits for most trace elements in refractory metal powders may be improved by adding a 'buffer' or 'carrier' to the powder sample. Based on the original work of Scribner and Mullin[3] various compounds have been proposed, such as gallium oxide and silver chloride. These additions promote fractional distillation of the more volatile impurities compared with the refractory matrix. This results in lower spectral complexity and better detection limits. For example, lead and bismuth can be determined in nickel alloys between 0.5 and 20 ppm using a mixed silver chloride-lithium fluoride buffer[4].

Detection limits may also be improved by sample pre-concentration methods, provided that the extra manipulations do not introduce unacceptable contamination. In an early paper by Balfour, Jukes and Thornton[5], several trace elements were separated from steel and nickel alloy matrices by co-precipitation of their sulphides using copper sulphide as a collector. The mixed sulphides were ignited to oxides at 500°C and analysed spectrographically by the graphite cup technique. The concentration ranges based on collection from a one gramme sample are given in Table 4.

In general, methods based on d.c. arc source optical emission spectroscopy have the advantage of multi-element capability and, where detection limits are low enough, provide a rapid procedure for quality control. However, they may be subject to severe interferences and matrix effects requiring large numbers of standard reference materials. Changes in the structure of materials, e.g. from cast to wrought, may introduce errors and for these reasons alternative methods are now generally preferred.

## 2. Optical Emission Spectroscopy using a Hollow Cathode Source

In the late 1960's it was becoming increasingly obvious that certain trace elements were affecting the properties of high-temperature materials and it was suspected that these effects were significant at levels below the detection limits of most readily available analytical techniques at that time. Following on from work by Mitchell and Harris[6], studies of the hollow cathode source were undertaken by Thornton[7]. These studies were directed at optimizing the technique for the determination of volatile trace elements in high-temperature alloys. As a result of this work commercial instrumentation became available in the early 1970's and this was later evaluated by Lowe[8] and Thelin[9].

It is fortuitous that most of the trace elements that cause severe problems in high-temperature materials are characterized by relatively high volatility. These include the most potent elements in terms of their effects on creep rupture properties, such as lead, bismuth, selenium and tellurium. We have seen previously that specifications for some nickel-base alloys now require that such elements are maintained below a few parts per million. The relatively high volatility of these elements compared with the principal alloying elements present in high-temperature materials has been used to great advantage in the hot hollow cathode source coupled to a conventional optical emission spectrometer.

In view of the importance of this technique to the aerospace industry - particularly within the United Kingdom - it is worth examining its performance in some detail.

A sample, in the form of chippings or powder, is weighed into a graphite crucible and inserted into the demountable hollow cathode lamp, which consists of a water-cooled stainless steel chamber, as illustrated diagrammatically in Figure 4. The lamp is evacuated and then continuously flushed with an inert gas, usually

82

helium, at a pressure of 10 to 20 Torr. A d.c. power supply of approximately 500 volts is applied between the lamp body (anode) and the graphite crucible (cathode) and under these conditions a suspended glow discharge is observed at the mouth of the graphite cathode. By increasing the power the cathode temperature is raised, causing melting of the sample and selective vaporization of the elements present. By careful choice of the operating conditions it is possible to obtain complete distillation of the more volatile trace impurity elements with minimal vaporization of the main alloying elements. This has considerable advantages in optical emission spectroscopy, leading to reduced spectral interferences and very low background levels.

The cathode is heated by bombardment with positively charged helium particles and can reach temperatures in excess of 2300°C. However, because of cooling effects caused by continuous gas flushing, the external wall temperature of the cathode may be considerably lower. This is illustrated in Table 5 which compares the external wall temperatures, measured by optical pyrometry, with the sample cavity temperatures, measured by material melting points, for a range of power inputs.

During operation the power is applied and held constant at a low value to remove residual atmospheric gases from the cathode. Integration then commences and the power is ramped linearly to a pre-set maximum value where it is held constant again. The variation of power with time is shown in Figure 5, together with a profile of the sample cavity temperature. As the graphite crucible temperature is raised, distillation and excitation of the elements present in the sample commences and, in general follows the order of their boiling points. Examples of typical curves for lead, tin and nickel are given in Figure 6. For most elements of interest the maximum power required is 400 watts (approximately 1 ampere, 400 volts d.c.) and under these conditions there is minimal vaporization of the main alloying elements found in high temperature materials. However, there are some exceptions to these generalizations.

Firstly, some of the trace elements do not appear in the discharge until much higher temperatures are reached than would be predicted from their individual boiling points. For instance, antimony has a boiling point of approximately 1750°C and is expected to vaporize at a lamp power of about 200 watts, whereas in nickel alloy matrices it does not distill completely below 2300°C (300 watts). Similar considerations apply to selenium and arsenic, suggesting that these elements are present in chemical combination in the alloy matrix. This leads to the interesting prospect of differentiation between 'free' and 'combined' element by time - resolved signal measurements. Of particular interest is the possibility of separately determining free and combined magnesium where this element is used to remove oxygen and sulphur.

Secondly, the matrix material may contain some alloying elements of relatively high volatility, such as manganese or copper. These will vaporize under normal operating conditions and may influence results in two ways. Volatile matrix elements may cause direct spectral overlap interference or they may have a quenching effect on the discharge, causing a reduction in analyte sensitivity. Such effects are uncommon, but may be of importance in the analyses of, for instance, high-manganese steels or cupro-nickels. Referring back to Figure 6, the most important requirement for optimization of operating parameters is to choose the lower and upper limits of the power (temperature) ramp to contain the total vaporization curve for the analyte and at the same time, minimum vaporization of unwanted matrix elements.

The separation of these distillation curves can be improved by careful selection of some of the other operating parameters, in particular the helium gas pressure and the electrode (cathode) geometry. In the former case there is only a narrow range of gas pressures for which stable discharges may be obtained in the lamp, but cathode geometry offers significant opportunities for optimizing individual element performance.

The cathode consists of a 6 mm diameter rod of high purity graphite drilled to a depth of 25 mm, into which the sample is weighed. Variation of the depth of the cathode cavity has a significant effect on analyte signal, as shown in Figure 7. In general, the most volatile elements show highest sensitivity from cavity depths in excess of 20 mm, which contrasts with the signal response of less volatile matrix elements such as iron, chromium and nickel. For most applications a compromise value of 25 mm has been chosen. The mechanism for these relationships probably involves two factors, namely increased sample temperature within the shorter cathodes and longer residence time of the vaporized analyte in the deeper cathodes.

The use of a commercially available hollow cathode source and a vacuum path optical emission spectrometer offers simultaneous analysis for a number of trace elements with exceptionally good limits of detection, as reported by Lowe[8]. Of fourteen elements investigated, he reports detection limits of less than 0.01 parts per million for four (cadmium, silver, thallium and magnesium), between 0.01 and 0.1 parts per million for eight (lead, bismuth, zinc, selenium, tellurium, indium, arsenic and antimony), with only tin (0.5 ppm) and gallium (2 ppm) above this level - see Table 6. Thelin[9] reported the determination of lead, bismuth, zinc, silver, antimony and calcium in a range of steels, nickel alloys and ferro-alloys with detection limits of less than 0.1 parts per million and a mean precision of better than 10% relative. He analysed several different materials using a single calibration for each element and reported excellent correlation with the certified values - this is illustrated for bismuth in Table 7. This work clearly identifies the sensitivity of the technique and its relative freedom from matrix effects.

Further examples of the sensitivity and applicability of this procedure are illustrated in Figures 8, 9 and 10, all of which are based on work performed in the author's laboratory. In Figure 8, values for tellurium are plotted for two nickel base casting alloys that were originally certified by neutron activation analysis in work carried out by Morris and Hill[10] at Brunel University. With an air path spectrometer using the line Te 214.2 nm the author obtained a detection limit of 0.1 parts per million.

Figure 9 demonstrates the sensitivity for the line Se 206.3 nm, again using an air path spectrometer. In this case the standard samples were of three nickel-base alloys analysed by three independent analytical procedures. A detection limit of about 1 part per million was obtained, but this has since been considerably improved by the use of a vacuum path spectrometer and the line Se 196.1 nm.

Finally, the correlation for Pb 283.3 nm is shown in Figure 10 for a variety of alloys, including steels and nickel-base alloys. In this case a correction for manganese was applied to overcome the lower sensitivity in the presence of up to two per cent of this element.

It is evident from the examples given that optical emission spectrometry using a hollow cathode source is capable of providing valuable information for volatile trace elements in high temperature materials. The technique simultaneously measures a number of elements, typically fourteen in nickel-base alloys, in about ten minutes. However, precision is not particularly impressive (generally 5-15% of content), but is satisfactory for most quality control applications. As with most techniques based on optical emission spectroscopy, the provision of suitable certified reference materials is often a problem.

## 3.    Methods based on Atomic Absorption Spectroscopy

The earliest exposition on the subject of atomic absorption spectroscopy was by Walsh[11] in 1955, after which there was little progress until about 1963. The popularity and use of this technique then underwent an explosive growth so that within a few years it had become established as one of the most important tools in analytical laboratories dealing with the determination of metals. It retains this position today.

Radiation from a primary light source, such as a hollow cathode lamp or an electrodeless discharge tube is passed through the sample vapour and selectively absorbed by free atoms of the analyte. A detector, usually a photomultiplier tube, is tuned to a characteristic wavelength of the light source corresponding with a suitable resonance wavelength of the analyte, and the absorption of the primary beam is measured. This is shown diagrammatically in Figure 11.

The primary beam may also undergo non-specific absorption due to the presence of 'smoke' particles in the light path and various compensation techniques are used to overcome those background effects. The most important are based on beam chopping using a deuterium arc continuum source, or Zeeman magnetic beam splitting[12]. Background correction becomes increasingly important as analyte concentrations decrease, particularly where measurements are carried out at short wavelengths and with non-flame atomizers.

Analyte vapour is generally produced by aspiration

of a solution of the sample into a flame or by resistively heating the sample in a graphite tube. For flame atomization the most commonly used fuel mixtures are air-acetylene and nitrous oxide-acetylene, producing temperatures of about 2200°C and 3000°C respectively. The more refractory elements are usually determined in the higher temperature nitrous oxide-acetylene flame, which also has some advantages in minimizing chemical interferences. However, where possible, the air-acetylene flame is preferred because of lower noise levels and better sensitivity for easily ionizable elements. An example of the use of flame atomic absorption spectroscopy is provided by the International Standards Organisation method, ISO 6351 Part 1[13], where nine elements are determined in nickel after dissolution in nitric acid. The concentration ranges are given in Table 8.

The ISO procedure is simple, direct and uses synthetic calibrations, but the sequential nature of the technique (i.e. one element at a time) can be a general problem in atomic absorption spectroscopy if many elements have to be determined. Using direct flame atomization, detection limits of approximately 1 part per million are obtainable for calcium, cadmium, lithium, magnesium, potassium, silver, sodium and zinc, based on a sample dilution of 10 g/l. A further group of elements, namely cobalt, chromium, copper, iron, manganese and nickel may be determined with detection limits of 2 to 5 parts per million. Although these detection limits may be satisfactory in some applications, it is often necessary to determine these and other elements at lower concentration levels. Flame techniques may be improved in this respect by separation and pre-concentration of the analyte(s) using methods such as solvent extraction, co-precipitation or ion exchange. For example, Thornton and Burke[14], determined lead, bismuth, tin and antimony in nickel-base alloys by extraction of the iodides of these analytes into a solution of tri-n-octylphosphine oxide (TOPO) in 4-methylpentan-2-one. The organic extract was aspirated directly into an air-acetylene flame and detection limits of 1 part per million or better were achieved for all four elements.

A specialized application of preconcentration has been reported by Thompson and Thomerson[15] for elements that form volatile hydrides. In this procedure, sodium borohydride is added to a solution containing the sample and the volatile hydrides are passed into an externally heated silica tube mounted along the optical axis of the instrument. They report detection limits for arsenic, bismuth, germanium, lead, antimony, selenium, tin and tellurium. Based on a sample dilution of 10 g/l and their recommended volume of 1 ml, the detection limits for solid samples are given in Table 9. However, Smith[16] found severe interference effects from a number of elements, notably nickel, cobalt, copper and iron. Subsequently, it has been shown that these interferences can be minimized, and in some cases eliminated by the addition of complexing agents to the sample solution. For instance, Drinkwater[17], has reported the determination of bismuth down to 0.1 parts per million in nickel alloys using EDTA as a complexing agent.

These specialized procedures are capable of achieving acceptably low detection limits, but they tend to be labour intensive, prone to contamination and of limited applicability. Of much more general interest is the use of resistively heated graphite 'furnaces' for the determination of trace elements.

In this technique, known variously as graphite furnace atomization (GFA) or electrothermal atomization (ETA), the sample is vaporized in a heated graphite tube rather than aspirated into a flame. The sample is unusually introduced into the GFA as a solution and subjected to a programmed heating cycle. This commonly consists of drying, ashing, atomization and cleaning stages, where the maximum temperature (normally reserved for atomization or cleaning) is about 2700°C and a typical duration for the complete cycle is 1-2 minutes. Very small volumes of sample solution are used, generally between 10 and 50 microlitres. The absolute sensitivity of this method is often very much better than flame atomic absorption, extending to the picogramme level for some elements. However, the limited volume of sample that can be introduced negates some of this advantage. Nevertheless, the technique is one of the most sensitive available in modern laboratories for many elements and has become the standard procedure for many applications, other than those involving the determination of refractory carbide forming elements such as niobium, tungsten, tantalum and zirconium. These elements form stable carbides in the graphite furnace and fail to volatilize.

A typical example of the application of this technique is described by Welcher, Kriege and Marks([18]), in which they determine lead, bismuth, selenium, tellurium and thallium in high-temperature alloys by dissolution in mixed nitric and hydrofluoric acids followed by direct atomization in a graphite tube furnace. Their recommended operating conditions, together with the detection limits achieved, are given in Table 10. This procedure, which subsequently became an 'Aerospace Recommended Practice' pulished by the Society of Automotive Engineers([19]), requires the use of a deuterium arc background corrector to eliminate broad band absorbence effects caused by particulate matter.

As so often happens, the continuing development of applications within this field has uncovered increasing problems and there is now a wealth of literature on aspects such as the chemical reactions occurring in the furnace, the use of 'matrix modifiers' to stabilize elements that would otherwise be lost during the ashing cycle and the use of different furnace geometries.

A similar technique has been extensively studied by Headridge and co-workers([20-23]), but in this case chippings or millings of the solid sample are dropped into a constant temperature inductively heated graphite furnace. Detection limits for volatile elements such as lead, bismuth, silver, selenium and thallium are generally better than 0.05 parts million. Although the equipment has not been commercially exploited, this work has led to the development of techniques for the introduction of chippings or millings of solid sample into the commercially available graphite furnace atomizers referred to previously. Some modifications, such as the use of platform inserts within the tube, have been necessary, but the equipment is essentially standard. Bäckman and Karlsson([24]) used this procedure to determine lead, bismuth, zinc, silver and antimony in a variety of steels and nickel base alloys. Their detection limits, based on the introduction of samples weighing a few milligrammes, are given in Table 11. They found that the results were influenced by the shape of the sample and that it was necessary to use independently analysed reference materials in the same form.

The value of atomic absorption spectroscopy as a technique for trace element determinations, particularly when used with graphite furnace atomization, cannot be over-emphasized and it will remain the preferred option for many laboratories on the grounds of simplicity, widespread applicability, sensitivity and cost.

## 4.   Inductively Coupled Plasma Spectrometry

The inductively coupled plasma is a high-temperature source which, when used in conjunction with an optical emission spectrometer, offers some advantages in trace element analysis, particularly for the determination of refractory elements. In this technique a solution of the sample is pumped to a nebulizer where it is converted to a fine aerosol, and subsequently injected into the central region of an inductively heated argon plasma, as shown schematically in Figure 12. The plasma is initiated with a high voltage spark, which ionizes some of the argon gas, and sustained with an electro-magnetic induction coil at frequencies of 25 MHz to 50 MHz. Temperatures as high as 9000°C may be obtained in the hottest region of the plasma, leading to complete chemical dissociation of the sample and vaporization of even the most refractory elements. At these high temperatures the emission spectrum may be very complex so that high quality spectrometers are needed to resolve the large numbers of spectral lines observed. Commercial systems are based on either sequential or simultaneous measurement of the analyte signals. In the former case a single detector traverses the spectrum measuring one element at a time. In the latter case detectors are pre-positioned for each analyte and these record the signals simultaneously. The choice of system is influenced by sample throughput, element programme, required flexibility and budget.

As a method for trace element analysis, it is most important to optimize the parameters that influence limits of detection for the analytes. These parameters include plasma power, observation height, choice of nebulizer and argon gas flows. For many elements the most important of these is the argon carrier gas flow to the nebulizer. This is demonstrated in Figure 13, which shows the variation in signal for the line Fe 259.9 nm for a range of carrier gas flow rates. The upper line represents the signal for a solution containing 2 µg/ml iron; the lower line is for an iron-free solution. In this case the best signal-to-noise ratio is obtained at a carrier gas flow rate of approximately 0.75 l/min. Optimum conditions vary from element to element, but it is usually possible to operate with a single set of good compromise conditions for a wide group of elements.

Using a single set of operating conditions and simultaneous measurement of the analytes, typical detection limits for some elements in steel are given in Table 12. These were obtained for solutions containing 2 g sample per litre. Performance may be improved by increasing the sample solution strength, use of alternative analyte wavelengths, or individual element optimization.

Most nickel base high-temperature alloys may be analysed using a single set of calibration curves by dissolution of the sample (2.5 grammes) in a mixture of hydrochloric (40 ml), nitric (10 ml) and hydrofluoric (5 ml) acids and subsequent dilution to 500 ml. Because of the presence of hydrofluoric acid, special nebulizers and torches must be used that are resistant to this solution. Detection limits for eighteen elements are given in Table 13. Of special interest are the detection limits for the refractory elements silicon, aluminium, molybdenum, niobium, hafnium, tantalum, titanium, tungsten and zirconium. The inductively coupled plasma procedure is particularly useful for these elements.

As with atomic absorption methods discussed previously, sample pre-treatment may be used to enrich the analyte prior to the use of inductively coupled plasma spectrometry. Such procedures include co-precipitation, ion exchange, solvent extraction and hydride generation. Because the sample is generally presented in the form of a solution, inductively coupled plasma techniques may be calibrated directly by the preparation of synthetic calibration standards. This is a major advantage compared with some of the solid sampling procedures, such as spark source OES or X-ray fluorescence, which rely on the availability of independently analysed reference standards.

## 5. Mass Spectrometric Methods

It has long been established that exceptionally low detection limits can be obtained for most elements by mass spectroscopic techniques, but early instrumentation was difficult to apply to the analysis of solids.

However, this measuring technique has recently been coupled to rapid and efficient ion sources and has resulted in the development of two new analytical techniques namely, inductively coupled plasma - mass spectrometry (ICP-MS), and glow discharge - mass spectrometry (GD-MS). The combination of these two ion sources with the inherent sensitivity of mass spectrometry has established the techniques as practical methods of analysis, capable of routine operation for the determination of most elements in the periodic table over wide concentration ranges.

Mass separation is based on the principle that when a charged particle (such as an ion) moves at right angles through a magnetic field, a force is exerted upon it. This causes the particle to move into a circular trajectory, the radius of which is proportional to the mass-charge ratio, m/e. An homogenous magnetic field can thus be used to separate ions of different masses which are then collected and measured with a device such as an electron multiplier.

The equipment used for ICP-MS is illustrated schematically in Figure 14.

In this example, a solution of the sample is aspirated into the ICP torch through a conventional nebulizer. The plasma impinges on a conical sampling port which has a small hole (0.5 to 1.0 mm) at its centre. The ions that pass through here impinge on a second conical (skimmer) port and then enter the body of the spectrometer. This two-stage sampling overcomes one of the major difficulties encountered in the development of ICP-MS, where it is necessary to reconcile the operating pressure of the ICP (essentially atmospheric) with the operating pressure of the spectrometer (approximately $10^{-6}$ Torr). The intermediate stage between the two ports is pumped down to about 1 Torr, allowing a successful combination of the two components.

Most of the characteristics of the ICP that have resulted in its significant growth as a source for optical emission spectroscopy are highly desirable in an ion source for mass spectrometry. It has a high ionization efficiency for most elements, resulting in almost complete conversion of analyte atoms to ions. At the same time, the energy is not sufficiently high to produce significant numbers of multiple-charged ions. This is important because of the overlap that occurs between singly charged ions of mass (M) with doubly charged ions of mass (2M). For example, overlap of $^{110}Cd^{++}$ on $^{55}Mn^+$ would be expected if doubly ionized cadmium is produced.

Although multiple charged ion production is limited in the ICP, interferences of this nature are present in the mass spectra, together with interferences caused by the presence of multicomponent ions. Examples of these are $OH^+$, $H_2O^+$, $ArH^+$, $NO^+$, $ClO^+$ and $Ar_2^+$. These effects are most noticeable in the low mass region and are influenced to some extent by the choice of acids used to dissolve the sample.

In a 1987 status report on ICP-MS, Selby and Hieftze[25] reported that, compared with ICP-OES, the technique suffers from some additional problems. These arise because of the possibility for interactions and reactions involving analyte ions in the intermediate (1 Torr) sampling region and because a cool surface is introduced into the plasma which can promote condensation and deposition. Because of this latter effect it is desirable to keep the level of dissolved solids relatively low, i.e. $\leqslant 0.2\%$ (m/v).

A comprehensive list of detection limits is given in Table 14, which is reproduced from one of the current manufacturers' literature. These values, which are quoted in terms of concentration of the analytes in solution, have the same numerical values as parts per million in a solid sample if we assume a sample solution strength of 0.1% (m/v). Because of some of the interference effects discussed previously, many of the detection limits may be significantly higher in real samples. It is interesting to note the relative uniformity of the detection limits. For the seventy-five elements where limits are quoted, all except twelve lie within the range 0.01 to 0.1 ng/ml.

In glow discharge - mass spectrometry (GD-MS) the principles of mass separation and detection are the same as those in ICP-MS. However, in this case the ICP ion source is replaced by a glow discharge ion source. This has the advantage of being capable of handling solid samples directly, without the necessity of taking them into solution.

An example of a glow discharge cell is shown in Figure 15. It consists of a tantalum chamber about 2 cm long and 1 cm in diameter filled with an inert gas (usually argon) at a pressure of about 0.5 Torr. When the sample is made the cathode and a few hundred volts applied between it and the chamber body, the gas breaks down and argon ions are produced. Under the applied potential difference, the argon ions are accelerated and collide with the surface of the sample. The impact of these collisions causes ejection of material from the sample surface by sputtering action. The sputtered material can take several forms, including positively charged ions and neutral species, but the ions are attracted back to the cathode. Neutral species diffuse into the glow discharge region of the plasma, are ionized, mainly by collision with metastable argon atoms, and subsequently pass into the mass spectrometer. As with the ICP source, the glow discharge source produces predominently singly charged ions with a high degree of efficiency, but some multiply charged atoms and charged molecular species are formed. Apart from ions arising from the sample, the spectrum will contain contributions from ionized species of the discharge gas and any atmospheric contaminants, but unlike ICP discharges, there are no difficulties encountered from the constituents of dissolving acids because the solid sample is analysed directly.

Sample presentation is usually in the form of small rods, about 2 mm in diameter, but adaptors are available for larger solids such as discs. Surface treatment to remove contaminants (by machining or chemical treatment) may be necessary, although the surface sputtering action of the glow discharge can be used to erode and clean the sample before commencement of the analysis. This gradual erosion of the surface can be used to perform depth profile analysis for contaminants on the surface of materials.

Detection limits for the glow discharge - mass spectrometry technique are similar to or better than those for ICP-MS, some examples being shown in Table 15.

In comparing the effectiveness of GD-MS with ICP-MS, the main factor to be considered is that the former uses solid sample excitation, whereas the latter (usually) requires the introduction of the sample in the form of a solution. This leads to both advantages and disadvantages. In particular the use of solutions is obviously best in those cases where the sample is supplied in that state, e.g. in the analysis of pickle acids or waste waters. It may also be more convenient to dissolve non-conducting solids for the ICP source, rather than mix them with a conducting powder for the GD source.  Similarly, samples of irregular form, shape or size, such as corrosion products or metal powders,may be more easily handled by dissolution.  However, the most important advantage of solution procedures is that they lend themselves readily to the preparation of synthetic calibration standards.  This may be a very significant advantage in the determination of trace elements in high temperature materials where, in general, there are very few certified reference materials.

Against these advantages one must consider the possible problems associated with dissolving a sample. For corrosion-resistant alloys this may be difficult without loss of some of the analytes through precipitation or volatilization.  Contamination from impurities in the acids and the laboratory environment may be significant and lead to unacceptably high blank levels.  As stated earlier, the acids themselves will contribute to the mass spectra with potential difficulties of spectral overlap.  It should also be considered that the act of dissolving a sample produces a dilution of the analytes which, for a 0.2% (m/v) solution represents a dilution factor of 500.  This suggests that solid sample presentation methods should be intrinsically more sensitive than solution procedures.

In concluding this section it should be noted that both of these mass spectrometric procedures may be used for the identification and measurement of individual element isotopes.  An example is shown in Figure 16, which clearly resolves the cadmium isotopes present in a solution containing one part per million of the element.

## CONCLUSIONS

In the last twenty years there has been a dramatic growth in the techniques available for determining trace elements in metals and alloys.  This has led to improved knowledge of the effects of trace elements and consequent proliferation within material specifications.

The techniques that have been described here represent, in the author's view, the most important developments in terms of their applicability on a wide basis.  It is recognized that whole disciplines have been omitted, such as atomic fluorescence spectrometry, electrochemical methods, X-ray fluorescence and photometric techniques.  In many laboratories these will have a significant role.

The inclusion of mass spectrometric techniques, although not widely used at present, is justified on the basis that their current state of development offers a significant opportunity for further understanding the role of trace elements in high-temperature materials.

## REFERENCES

1.  G. Mayer and C.A. Clark: Metallurgist and Materials Technologist, (1974), 491.

2.  Methods for Emission Spectrochemical Analysis, Method E-2 SM 8-21, published by American Society for Testing and Materials.

3.  B.F. Scibner and H.R. Mullin: J. Nat. Bur. Standards, 37 (1946), 379.

4.  Methods for Emission Spectrochemical Analysis, Method E483-74, published by American Society for Testing and Materials.

5.  B.E. Balfour, D. Jukes and K. Thornton: App. Spec., 20 (1966), 168.

6.  G.P. Mitchell and C.I. Harris: Proc. Soc. Analyt. Chem., (1965) 105.

7.  K. Thornton: Analyst, 94 (1969), 958.

8.  D.S. Lowe: Analyst, 110 (1985), 583.

9.  B. Thelin: App. Spec., 35 (1981), 303.

10. D.F. Morris and N. Hill: Metallurgia, 424 (1965), 99.

11. A. Walsh: Spectrochim Acta, 7 (1955), 108.

12. E. Grassam and J.B. Dawson: Europe. Spec. News, 21 (1978), 27.

13. International Standards Organisations, ISO 6351 (1985).

14. K. Thornton and K.E. Burke; Analyst, 99 (1974), 469.

15. K.C. Thompson and D.R. Thomerson: Analyst, 99 (1974), 595.

16. A.E. Smith: Analyst, 100 (1975), 300.

17. J.E. Drinkwater: Analyst, 101 (1976), 672.

18. G.G. Welcher, O.H. Kriege and J.Y. Marks: Anal. Chem., 46 (1974), 1227.

19. Aerospace Recommended Practice, ARP 1313, published by Society of Automotive Engineers (1975).

20. J.B. Headridge, J.B. and D.R. Smith: Talanta, 19 (1972), 833.

21. J.B. Headridge and R. Thompson: Anal. Chim. Acta, 102 (1978), 33.

22. A.M. Asiz-Alrahman and J.B. Headridge: Analyst, 104 (1979), 944.

23. A.A. Baker, J.B. Headridge and R.A. Nicholson: Anal. Chim. Acta, 113 (1980), 47.

24. S. Backman and R.W. Karlsson: Analyst, 104 (1979), 1017.

25. M. Slby and G.M. Hieftje: Int. Laboratory, Oct. 1987, 28.

Table 1

**Specification maxima for trace metals in cast superalloys, ppm**

|               | Ag | Zn  | Cd  | Ga  | In  | Tl  | Sn  | Pb | As  | Sb  | Bi  | Se  | Te  |
|---------------|----|-----|-----|-----|-----|-----|-----|----|-----|-----|-----|-----|-----|
| Prior to 1975 | 5  | ... | ... | ... | ... | ... | ... | 10 | ... | ... | 1   | ... | ... |
| AMS 2280*     | 50 | 50  | 50  | 50  | 50  | 5   | 50  | 5  | 50  | 50  | 0.5 | 3   | 0.5 |
| Current       | 5  | 5   | 0.2 | 30  | 0.2 | 0.2 | 30  | 5  | 30  | 3   | 0.5 | 1   | 0.5 |

*Aerospace Material Specification, AMS 2280 (1975)
(Total must not exceed 400 ppm)

Table 2

**Detection limits for 80% nickel, 20% chromium alloy by
point-to-plane d.c. arc spectrography**

| Element             | Ag | B  | Si | Zr | Sn | Pb | Ce | Sb | Nb  | As  |
|---------------------|----|----|----|----|----|----|----|----|-----|-----|
| Detection limit, ppm| 1  | 10 | 10 | 10 | 20 | 5  | 50 | 50 | 100 | 150 |

## Table 3

**Concentration ranges for spectrographic analysis of molybdenum
by d.c. arc spectrography**

| Element | Concentration range, ppm |
|---|---|
| Aluminium | 10 to 200 |
| Iron | 5 to 200 |
| Tin | 3 to 200 |
| Silicon | 3 to 200 |
| Nickel | 1 to 200 |
| Chromium | 3 to 100 |
| Lead | 2 to 100 |
| Manganese | 1 to 50 |
| Copper | 2 to 50 |
| Calcium | 0.5 to 30 |
| Magnesium | 0.1 to 5 |

ASTM Suggested Procedure, E-2 SM 8-21

## Table 4

### Concentration ranges for d.c. arc spectrography after pre-concentration on copper sulphide

| Element | Concentration range, ppm |
|---------|--------------------------|
| Silver | 0.1 to 10 |
| Arsenic | 50 to 1000 |
| Gold | 5 to 100 |
| Bismuth | 0.5 to 25 |
| Cadmium | 1 to 25 |
| Indium | 1 to 100 |
| Lead | 0.5 to 50 |
| Antimony | 5 to 500 |
| Tin | 0.5 to 50 |
| Thallium | 1 to 100 |

Elements collected from 1g sample on to 0.15g copper as mixed sulphides.

Table 5

**Temperature of graphite hollow cathode**

| Power (watts) | Cathode lip (°C) | Sample cavity (°C) |
|---|---|---|
| 100 | 1280 | 1130 |
| 250 | 1430 | 1820 |
| 400 | 1580 | 2330 |

Hollow cathode lamp operated in a flushing helium atmosphere at 15 Torr.

Table 6

**Limits of detection for nickel alloy analysis by hollow cathode - OES**

| Element | Wavelength (nm) | Detection limit, ppm |
|---|---|---|
| Lead | 283.3 | 0.03 |
| Bismuth | 306.7 | 0.01 |
| Tin | 317.5 | 0.5 |
| Cadmium | 326.1 | 0.001 |
| Silver | 328.0 | 0.001 |
| Zinc | 334.5 | 0.1 |
| Thallium | 351.9 | 0.004 |
| Selenium | 196.1 | 0.1 |
| Gallium | 417.2 | 2 |
| Tellurium | 214.2 | 0.06 |
| Indium | 451.1 | 0.02 |
| Arsenic | 234.9 | 0.18 |
| Antimony | 259.8 | 0.1 |
| Magnesium | 285.2 | 0.001 |

Results obtained using a commercial hollow cathode source unit and a vacuum path direct-reading spectrometer.
Sample size: 10 mg.

## Table 7

### Determination of bismuth in steels, ferroalloys and nickel alloys by hollow cathode - OES

| Sample | Grade | Reference value (ppm) | HCS-OES value (ppm) |
|--------|-------|----------------------|---------------------|
| STEELS | | | |
| BCS 331 | 15 Cr/6 Ni | 0.05 | 0.05 |
| BCS 336 | 18 Cr/9 Ni | 3.4 | 3.5 |
| NBS 361 | 1 Cr/2 Ni | 6.0 | 6.4 |
| NBS 364 | 0.1 Cr/0.1 Ni | 24. | 24. |
| BCS 334 | 26 Cr/21 Ni | 0.05 | 0.05 |
| NICKEL ALLOYS | | | |
| D5-911 | 12 Cr/69 Ni/8.5 Co | 0.58 | 0.58 |
| D5-912 | " | 1.18 | 1.17 |
| FERRO ALLOYS | | | |
| 1034/4-24 | Ferromolybdenum | 2.8 | 2.9 |
| 1138/2-27 | " | 6.6 | 6.9 |
| 1137/2-34 | " | 10. | 9.5 |

## Table 8

### Concentration ranges for analysis of nickel by flame atomic absorption spectrometry

| Element | Concentration range, ppm |
|---------|--------------------------|
| Silver | 2 to 100 |
| Bismuth | 10 to 100 |
| Cadmium | 2 to 25 |
| Cobalt | 10 to 100 |
| Copper | 2 to 100 |
| Iron | 25 to 100 |
| Manganese | 5 to 100 |
| Lead | 5 to 100 |
| Zinc | 2 to 25 |

International Standards Organisation, ISO 6351 (1985).

## Table 9

### Detection limits for elements determined as their volatile hydrides by atomic absorption spectrometry

| Element | As | Bi | Pb | Sb | Se | Sn | Te |
|---|---|---|---|---|---|---|---|
| Detection limited (µg/g) | 0.12 | 0.03 | 15 | 0.08 | 0.27 | 0.08 | 0.21 |

The above limits are calculated from LOD's in pure analyte solutions, as published by Thompson and Thomerson. See text for discussion of possible interferences.

## Table 10

### Typical operating conditions and detection limits for the analysis of nickel alloys by GFA-AAS

| Element | Pb | Bi | Te | Se | Tl |
|---|---|---|---|---|---|
| Wavelength nm | 283.3 | 223.1 | 214.3 | 196.0 | 276.9 |
| Solution volume, µl | 50 | 20 | 50 | 50 | 50 |
| Sample concentration, mg/ml | 20 | 20 | 20 | 20 | 20 |
| Drying temperature, °C | 150 | 150 | 150 | 600 | 150 |
| Ashing temperature, °C | 400 | 800 | 600 | 1000 | 500 |
| Atomizing temperature, °C | 2000 | 2200 | 2200 | 2400 | 2000 |
| Detection limit, ppm* | 0.1 | 0.1 | 0.2 | 0.1 | 0.1 |

*Deuterium arc background correction used.

## Table 11

### Detection limits for the analysis of steels and nickel alloys by GFA-AAS using direct insertion of solid samples

| Element | Pb | Bi | Sb | Zn | Ag |
|---|---|---|---|---|---|
| Detection limit, ppm | 0.02 | 0.03 | 5 | 1 | 0.01 |

*Deuterium arc background correction used.

103

## Table 12

## Typical limits of detection for some elements in steel by ICP-OES

| Element | Wavelength, nm | Detection limit, ppm |
|---------|----------------|----------------------|
| Silicon | 251.6 | 22 |
| Manganese | 257.6 | 0.2 |
| Cobalt | 228.6 | 4 |
| Chromium | 284.9 | 10 |
| Molybdenum | 202.0 | 4 |
| Niobium | 295.0 | 8 |
| Vanadium | 290.8 | 5 |
| Zirconium | 343.8 | 4 |
| Tin | 189.9 | 8 |

The above values were obtained on solutions containing 2g sample/litre and compromise operating conditions for simultaneous measurement.

## Table 13

### Typical detection limits for some elements in nickel alloys by ICP-OES

| Element | Wavelength, nm | Detection limit, ppm |
|---|---|---|
| Silicon | 251.6 | 3 |
| Manganese | 257.6 | $<1$ |
| Boron | 182.5 | 6 |
| Aluminium | 396.1 | 4 |
| Calcium | 393.3 | $<1$ |
| Cobalt | 238.8 | 2 |
| Chromium | 284.9 | 5 |
| Copper | 324.7 | 2 |
| Iron | 259.9 | 1 |
| Magnesium | 279.5 | $<1$ |
| Molybdenum | 202.0 | 2 |
| Niobium | 295.0 | 3 |
| Hafnium | 339.9 | 4 |
| Tantalum | 268.5 | 4 |
| Titanium | 368.5 | 1 |
| Vanadium | 290.8 | 1 |
| Tungsten | 207.9 | 11 |
| Zirconium | 343.8 | 1 |

The above values were obtained for solutions containing 5g sample/litre dissolved in mixed hydrochloric, nitric and hydrofluoric acids.

## Table 14
## Detection limits for ICP-MS

| IA | IIA | IIIB | IVB | VB | VIB | VIIB | VIII | VIII | VIII | IB | IIB | IIIA | IVA | VA | VIA | VIIA | O |
|---|---|---|---|---|---|---|---|---|---|---|---|---|---|---|---|---|---|
| H | | | | | | | | | | | | | | | | | He |
| 0.06 Li | 0.1 Be | | | | | | | | | | | 0.08 B | 50 C | N | O | 30* F | Ne |
| 0.06 Na | 0.10 Mg | | | | | | | | | | | 0.1 Al | 10 Si | 2* P | S | Cl | Ar |
| 1* K | 5 Ca | 0.08 Sc | 0.06 Ti | 0.03 V | 0.02 Cr | 0.04 Mn | 0.2 Fe | 0.01 Co | 0.03 Ni | 0.03 Cu | 0.08 Zn | 0.08 Ga | 0.08 Ge | 0.02 As | 0.04 Se | 0.01 Br | Kr |
| 0.02 Rb | 0.02 Sr | 0.01 Y | 0.03 Zr | 0.02 Nb | 0.08 Mo | Tc | 0.05 Ru | 0.02 Rh | 0.06 Pd | 0.04 Ag | 0.07 Cd | 0.01 In | 0.03 Sn | 0.02 Sb | 0.04 Te | 0.01 I | Xe |
| 0.02 Cs | 0.02 Ba | 0.01 La | 0.03 Hf | 0.02 Ta | 0.06 W | 0.01 Re | 0.01 Os | 0.06 Ir | 0.08 Pt | 0.08 Au | 0.08 Hg | 0.05 Tl | 0.02 Pb | 0.06 Bi | Po | At | Rn |
| Fr | Ra | Ac | | | | | | | | | | | | | | | |

| | | | | | | | | | | | | | |
|---|---|---|---|---|---|---|---|---|---|---|---|---|---|
| 0.01 Ce | 0.01 Pr | 0.01 Nd | Pm | 0.01 Sm | 0.02 Eu | 0.04 Gd | 0.01 Tb | 0.04 Dy | 0.01 Ho | 0.02 Er | 0.01 Tm | 0.03 Yb | 0.01 Lu |
| 0.02 Th | 0.02 Pa | 0.02 U | Np | Pu | Am | Cm | Bk | Cf | Es | Fm | Md | No | Lw |

Values are quoted as µg/l in solution

**Table 15**

**Typical limits of detection for elements in
nickel alloys by glow discharge - mass spectrometry**

| Element | LOD, ppm | Element | LOD, ppm |
|---|---|---|---|
| Cobalt | 0.01 | Lead | 0.01 |
| Tungsten | 0.01 | Bismuth | 0.004 |
| Aluminium | 0.01 | Silver | 0.008 |
| Niobium | 0.01 | Selenium | 0.01 |
| Tantalum | 0.01 | Tellurium | 0.006 |
| Vanadium | 0.01 | Thallium | 0.01 |
| Copper | 0.01 | Tin | 0.01 |
| Hafnium | 0.01 | Indium | 0.004 |
| Zirconium | 0.01 | Antimony | 0.006 |
| Titanium | 0.01 | Arsenic | 0.01 |

## FIGURE 1

### ILLUSTRATIONS OF GOOD AND POOR DETECTABILITY

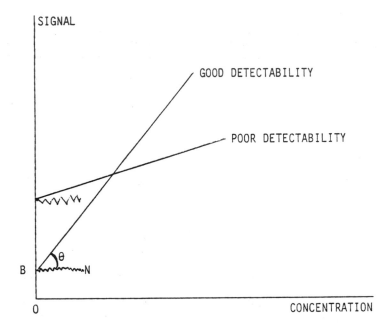

Limit of detection improves with decreasing background (B), increasing sensitivity (signal per unit concentration, θ) and decreasing noise level (N).

## FIGURE 2

**RELATIONSHIP BETWEEN ANALYTE CONCENTRATION AND
PRECISION AT OR NEAR THE DETECTION LIMIT**

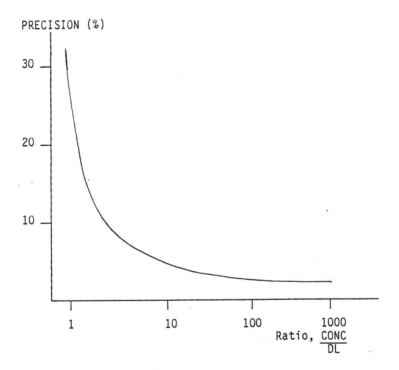

At the delection limit (DL, the precision is ±33%; at
3 x DL the precision is +11%.
Precision normally improves to some limiting value at
concentrations above 100 x DL.

FIGURE 3

ELECTRODE CONFIGURATIONS FOR d.c. ARC
OPTICAL EMISSION SPECTROSCOPY

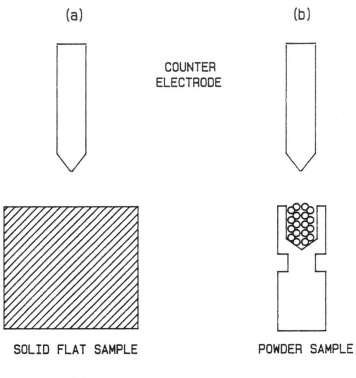

(a)     (b)

COUNTER
ELECTRODE

SOLID FLAT SAMPLE          POWDER SAMPLE

(a) Point-to-Plane configuration
(b) Graphite Cup Electrode

# FIGURE 4

## DEMOUNTABLE HOLLOW CATHODE LAMP

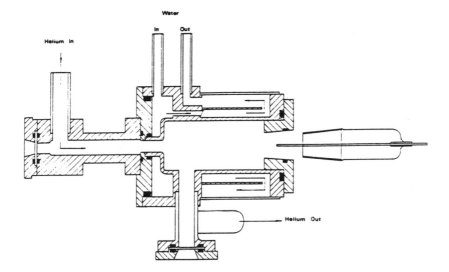

## FIGURE 5

## TYPICAL POWER AND TEMPERATURE RAMPS FOR
## HOT HOLLOW CATHODE SOURCE

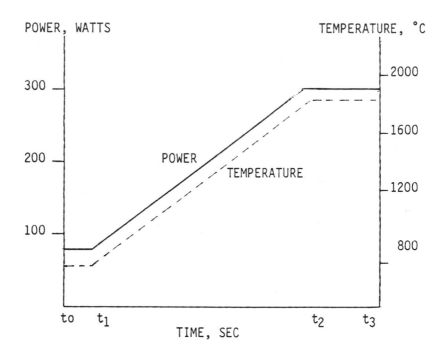

$t_0 - t_1$ : pre-integration (30 sec at 60 watts)
$t_1 - t_2$ : power-ramp (200 sec at 1.2 watts/sec)
$t_2 - t_3$ : constant power (30 sec at 300 watts)

## FIGURE 6

**TYPICAL DISTILLATION CURVES FOR
HOT HOLLOW CATHODE SAMPLE**

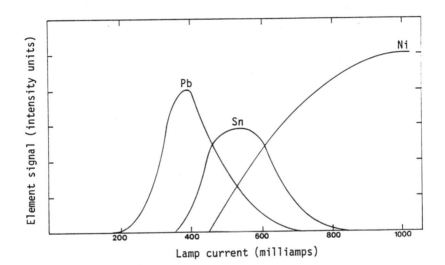

FIGURE 7

RELATIONSHIP BETWEEN SIGNAL INTENSITY AND DEPTH OF
GRAPHITE CATHODE FOR HOT HOLLOW CATHODE LAMP

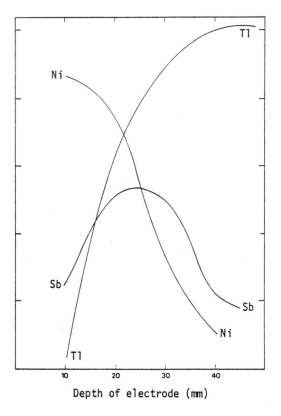

Depth of electrode (mm)

## FIGURE 8

**CALIBRATION GRAPH FOR TELLURIUM BY
HOLLOW CATHODE SPECTROSCOPY**

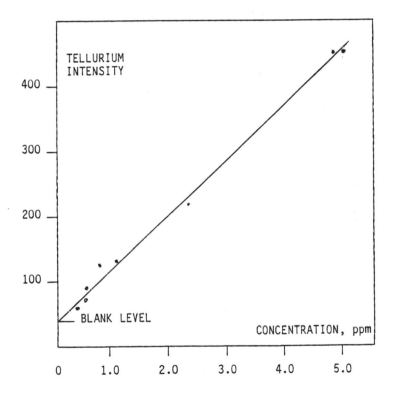

Tellurium reference values determined by neutron
activation analysis

FIGURE 9

CALIBRATION GRAPH FOR SELENIUM BY
HOLLOW CATHODE SPECTROSCOPY

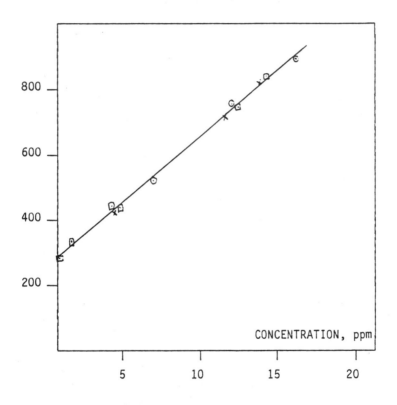

Selenium reference values determined by the
following techniques:-

⊖   Average values of AAS (hydride),
    AAS (graphite furnace) and X-ray fluorescence

x   AAS (graphite furnace)

⊡   AAS (hydride)

FIGURE 10

**CALIBRATION GRAPH FOR LEAD BY
HOLLOW CATHODE SPECTROSCOPY**

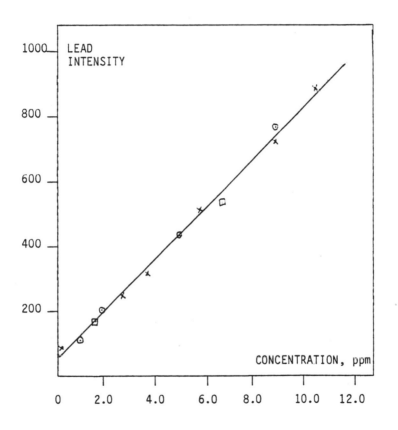

Hollow cathode intensities have been corrected for
manganese interference

⊘  Nickel alloys (< 0.2% Mn); x steels
   (up to 1.85% Mn)

◻  35 Ni-45 Fe-20 Cr alloy (1% Mn)

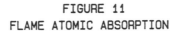

FIGURE 11
FLAME ATOMIC ABSORPTION

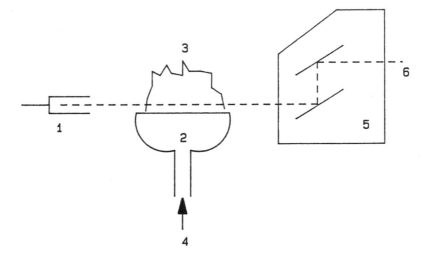

1) Light source  2) Burner head  3) Gas flame
4) Sample inlet tube  5) Monochromator
6) Light to detector

FIGURE 12

INDUCTIVELY  COUPLED  PLASMA

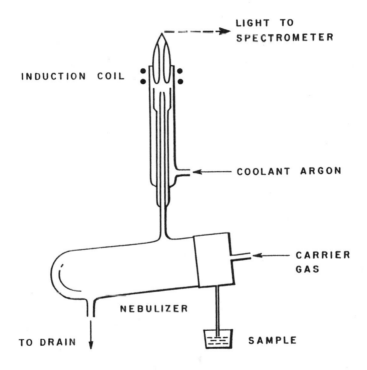

FIGURE 13

**EFFECTS OF CARRIER GAS IN INDUCTIVELY COUPLED PLASMA
- OPTICAL EMISSION SPECTROSCOPY**

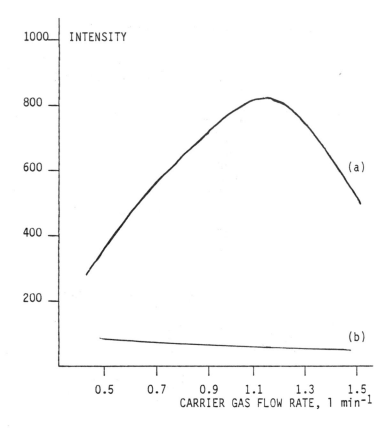

(a)  Solution containing 2µg/ml iron
(b)  Blank solution

Measurements made using a cross flow nebulizer on
the line Fe 259.9 nm

FIGURE 14

SCHEMATIC DIAGRAM OF ICP - MASS SPECTROMETRY SYSTEM

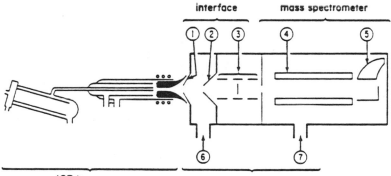

1. Sampling cone;  2. Skimming cone;  3.  Ion optic;
4. Mass spectrometer; 5. Detector; 6. Vacuum (1 Torr);
7. Vacuum ($10^{-5}$)

Reproduced from Perkin Elmer literature.

FIGURE 15

EXAMPLE OF GLOW DISCHARGE CELL FOR GD - MASS SPECTROMETRY

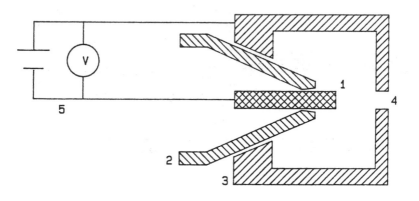

1) Rod-shaped conducting sample        2) PTFE sample holder
3) Tantalum glow discharge cell        4) Ion exit slit
5) Glow discharge power supply

## FIGURE 16

### MASS SPECTRUM OF A SOLUTION CONTAINING
### 1 ppm  CADMIUM USING AN ICP ION SOURCE

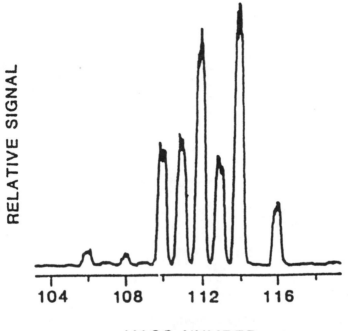

The individual cadmium isotopes are clearly resolved

# 4 : Energy dispersive spectroscopy
# K McKINDLEY

Kathryn McKindley is in the Department of Wrought
Technology and Process Physics, at Rolls-Royce plc,
P.O. Box 31, Derby.

## 1  INTRODUCTION

The properties of high-temperature materials are
strongly dependent on their chemical composition. Such
materials as nickel-based superalloys, high-strength
steels, and more recently ceramics and metal matrix
composite materials all pose a challenge to the
materials scientist in determining the effect of
composition and processing on their microstructure. In
the development of nickel-based superalloys, for
example, the use of high magnification electron optical
imaging has enabled the metallurgist to see the tiny
coherent precipitate phases (called gamma prime) which
give the alloys their inherent strength. Having
identified such phases the next step in the process of
material characterisation is to analyse them to
determine their composition.

The most common analysis technique which is able
to combine high magnification imaging and chemical
analysis is that of X-ray microanalysis. This
technique concentrates on the analysis of X-rays
emitted from a sample when it is bombarded with high-
energy electrons, in a vacuum, as is the case in the
electron microscope. It was not long after X-rays were
first discovered in 1895 that it became evident that

124

X-ray energies were intimately related to the atomic structure of the elements that emit them. When an electron beam is used to bombard a small area of a solid specimen a number of interactions occur resulting in a variety of signals for analysis (fig.1). These include the emission of characteristic X-rays superimposed on a background of continuous X-radiation (or Bremsstrahlung). As each element, when stimulated to do so, emits a unique pattern of X-ray energies, the elements in the sample can easily be identified and from the relative intensities of the characteristic lines emitted from the different elements in the sample the composition can be calculated.

In a solid target the X-ray spectrum originates from a volume of a few cubic micrometers below the surface, hence the term 'microanalysis',whereas in a thin sample this volume, due to reduced electron scattering, is smaller providing a resolution of about 50 nanometers.

As these X-rays are electromagnetic radiation they can be thought of as packets of energy called photons. The wavelength of the X-ray photons is related to the energy by Planck's equation:

$$\lambda = hc/E$$

where $\lambda$ is the wavelength of the X-ray, c is the speed of light, h is Planck's constant and E is the energy of the X-ray.

It can be seen therefore that an X-ray can be characterised by either its wavelength or energy. This has given rise to two methods of detecting the emitted X-ray spectrum from a sample by;-

(a)    a crystal diffracting spectrometer which detects the X-rays by measuring their wavelength. This is known as wavelength dispersive spectrometry (WDS) and will be dealt with in more detail in the next chapter. Figure 2 illustrates the detection of X-rays by a wavelength dispersive spectrometer along with a typical WDS spectrum obtained from a stainless steel sample.

(b)    a solid state detector which measures the X-ray energy known as energy dispersive spectrometry (EDS) (see fig.3).

This chapter concentrates on the energy dispersive system of X-ray analysis. It aims to give the reader a broad outline of the major principles and applications of the technique with particular emphasis on its application to the characterisation of high-temperature materials. For a full appreciation of X-ray analysis however it is important to have an understanding of the advantages and disadvantages of both EDS and WDS in terms of their performance to provide qualitative and quantitative analysis.

The addition of X-ray analysis facilities to the scanning electron microscope (SEM), transmission electron microscope (TEM) and the electron microprobe has offered an invaluable analytical facility to the electron microscopist which does not interfere with normal microscope operations. Together they provide powerful tools for the materials scientist to answer the questions of;

'What does it look like ?' and

'What is it made from?'

at micron and sub-micron resolution.

The development in recent years of the Si(Li) solid state energy resolving spectrometer has revolutionised X-ray microanalysis. This detector does away with the complexities of the diffracting crystal and physically dispersed spectrum of X-rays as used in WDS. Instead the energy dispersive (ED) system collects emitted X-rays of all energies and sorts them electronically to provide a simultaneous recording of the whole X-ray energy spectrum emitted from the sample (fig.3).

The solid state ED detector provides an efficient system for elemental analysis and due to its speed and simplicity it is now the most common X-ray measurement instrument found on the electron microscope.

One area where EDS falls down with respect to WDS is in spectral resolution. Figures 2(b) and 3(b) compare the spectra obtained from a stainless steel sample using WDS and EDS respectively. It is seen that the spectral lines are much broarder in the ED spectrum than they are in the WD spectrum. This means that in EDS there is a greater problem with overlapping peaks.

In this cursory treatment of ED we shall begin by explaining the primary results of the interaction of an electron beam with a sample in which the emission of X-rays are of primary interest. Then we shall consider how these X-rays are detected and counted by the ED spectrometer to produce data which can provide qualitative and quantitative results.

## 2  ELECTRON INTERACTIONS IN THE SAMPLE

As X-ray microanalysis is used in conjunction with normal electron microscopic techniques it is important that the user has an appreciation of the principal results of the interaction of an electron beam with a specimen and the physics of the processes which occur.

In the electron column, electrons are accelerated through an electric field thus gaining kinetic energy. Typical accelerating voltages are 10-30kV for SEMs and 100-400kV for TEMs providing electrons in the energy ranges of 10-30keV and 100-400keV respectively. When these primary electrons impinge on a sample they impart this energy via a series of elastic and inelastic collisions until they finally come to rest at a certain distance beneath the surface. This gives rise to number of signals being emitted from the sample including characteristic X-rays. Figure 4 illustrates the reaction volumes for the various interactions showing the source of the different signals. This distribution is often referred to as the 'teardrop'. In a thin sample however the lateral diffusion of electrons is reduced as the signals are produced by interactions at the top of the teardrop (fig.4).

To explain the physics of these processes we will use the Bohr model of the atom which assumes the atom is comprised of a positively charged nucleus surrounded by electrons in discrete orbitals at different energy levels.

### 2.1 BACKGROUND X-RADIATION

As the electrons entering the sample decelerate due to inelastic collisions with the atoms in the sample, background X-rays are produced (also known as the continuum, white radiation or Bremsstrahlung - which is German for 'braking radiation'). As the electrons decelerate, background X-rays are generated whose energies range from zero to the energy of the incoming electrons (fig.5(a)). The energy of X-rays produced depends on the amount of energy the incoming electron loses when it collides with an atom. For example if in a collision an electron lost all of its energy the resulting X-ray energy will be equal to that of the incoming electron. This is the maximum energy that an X-ray can have. Figure 6 contains X-ray energy spectra obtained from a pure nickel sample using an accelerating voltage of 5,10 and 20kV. It is seen that the background radiation has a sharp cut-off at the accelerating voltage, $E_o$, used (remember an accelerating voltage of 10kV results in primary electrons of energy 10keV). A fall-off at low energies is seen due to X-ray absorption between the X-ray source and the detector.

## 2.2 CHARACTERISTIC X-RAYS

An incoming, high-energy electron can interact with the inner shell of an atom resulting in ejecting an electron from a particular shell or orbital. This has converted the atom, in a stable energy state, into an unstable ion at a higher energy level. To return to its original energy state, or ground state, the ion has to give up some energy. The most likely way in which it can do this is for a series of transitions to take place where an electron from a higher energy orbital falls to occupy the vacated position in the inner shell (fig.5(a)). This process reduces the potential energy of the ion thus releasing the excess energy as an X-ray photon. The energy of the X-ray therefore is determined by the difference in energy between the two energy levels. Each element therefore, when excited to do so, will emit a unique pattern of X-ray lines according to its atomic structure.

The X-ray lines are named according to the shell in which the initial vacancy occurred and the shell from which an electron falls to fill the vacancy. For example when an electron is ejected from the innermost K shell, the subsequent X-ray lines produced are known as K lines. If the line was produced due to an electron from the next shell occupying the vacancy, in this case the L shell, the line would be called the Kα line. If the vacancy had been filled by an electron from two orbitals away it would be called a Kß line and so on. Figure 5(b) illustrates the production of characteristic X-rays from different energy levels within an atom. As the atomic number or the atom increases so does the number of possible transitions due to the increased number of orbitals. For example oxygen (Z=8) has two electrons in the K shell and 6 in the L shell. Therefore the only X-ray line that can be generated is the Kα line as there are no electrons in the M shell to form the Kß or Lα lines. This also means that with increasing atomic number the more complex the X-ray spectrum. In X-ray microanalysis we are usually concerned with the K,L and M-series of X-rays which are referred to as KLM lines.

Characteristic X-rays can only be generated if the incoming electrons have sufficient energy to remove one of the inner core electrons. The energy required to

128

produce such an ionised state will always be larger than the energy of any subsequent electron transition. The minimum energy required to remove an electron from its shell is known as the critical excitation energy ($E_c$) and has a discrete value for each shell and subshell. It is seen in fig.6 that the characteristic X-rays for nickel do not appear until the accelerating voltage is 10kV. This is because the critical excitation energy for the nickel K shell is 8.33keV. Within a particular family of X-ray lines the line which will be most efficiently excited will therefore depend on the accelerating voltage of the microscope. It is seen in fig.6 that although the NiKα and Kß lines appear using an accelerating voltage of 10kV they are not as efficiently excited as when using 20kV. To efficiently excite any X-ray line, an accelerating voltage of at least 2.5 times the energy of the X-ray is required.

## 2.3 SECONDARY ELECTRONS
The most popular signal used for electron imaging is the secondary electron. These are produced when a primary electron imparts some of its energy to a weakly bound electron in the sample thus giving it enough energy to escape. The secondary electrons have low energies (<50eV) so only those close to the surface have sufficient energy to escape. The emission of secondary electrons is therefore very sensitive to topography as electrons have a greater probability of escaping from a peak than a hole (fig.7(a)). If the primary beam is scanned across the surface of the sample and the intensity of the secondary X-rays is detected and used to modulate the brightness of a spot on a video monitor, a secondary electron image showing the topography of the sample is formed. This image carries little chemical information but its sensitivity to topography and its good spatial resolution makes it the most popular choice of micrographic imaging.

## 2.4 BACKSCATTERED ELECTRONS
When primary electrons interact with the nucleus of an atom they can be backscattered out of the sample with little energy loss (fig.7(b)). As the backscattered electrons (BSE) have a much larger energy than secondary electrons, they can escape from much greater depths in the sample. These electrons therefore do not carry as much topographical information as secondary electrons, nor do they have as good resolution, but are influenced by the mean atomic number of the specimen.

The higher the atomic number of the elements present the greater the BSE signal. This is because the size of the positively charged nucleus increases with atomic number so there is a greater possibility of the electrons to interact.

## 2.5 AUGER ELECTRONS

De-excitation of an ionised atom can occur via an alternative process to the emission of characteristic X-rays. When an inner shell vacancy is filled by an outer electron, as just described, the X-ray produced can be reabsorbed by the atom resulting in the emission of a lower energy electron (fig.7(c)). These ejected electrons are called Auger electrons and will possess an energy equal to to the difference in the original characteristic X-ray and the binding energy of the ejected electron. The energy of these Auger electrons contains specific chemical information about the atom from which it originated. As the Auger electrons have low energy they can only escape from the first few atomic layers of the sample.

The probability of whether a characteristic X-ray or an Auger electron is emitted from the sample is dependent primarily on the atomic number of the excited atom. This is referred to as the fluorescent yield. Elements with low atomic numbers have a low fluorescent yield and the emission of Auger electrons is favoured, whereas at higher atomic number the emission of characteristic X-rays is preferred.

## 2.6 SECONDARY FLUORESCENCE

As X-rays pass through the sample on their way out they may interfere with with other atoms causing further X-ray emissions. This is called X-ray fluorescence and its effect can be that the intensity of primary characteristic X-ray lines can be either enhanced or reduced. For fluorescence to occur, the radiation giving rise to it must understandably have a higher energy than the secondary radiation produced.

The secondary X-rays caused by fluorescence cannot be distinguished from those generated by the primary electron beam so this can cause problems when doing quantitative analysis (see 3.3). As the volume of material which is excited via secondary fluorescence is larger than that of the primary X-rays, when analysing near to interfaces and boundaries it is possible to obtain misleading results due to material being excited via fluorescence at the other side of a boundary.

# 3 X-RAY INTENSITIES

The intensities of the emitted X-ray lines are
strongly affected by both the operating conditions of
the microscope and the composition of the sample. For
quantitative analysis the concentrations of the
constituent elements in the sample are calculated, in
either atomic or weight percent, from the measured
intensities of the characteristic X-ray lines. To do
this the intensities of the X-ray peaks are usually
compared to those obtained from a pure element standard
or a standard of known composition using identical
conditions. Ideally the ratio of a particular X-ray
line emitted from the sample (I'), to that emitted from
a standard of known concentration ($C_o$) is proportional
to the concentration of the element in the sample (C')
according to:

$$C' = C_o I'/I_o$$

For example under the same operating conditions a
sample containing 50% nickel should radiate half as
many nickel X-rays as a pure nickel sample. This
however is only an approximation because of the complex
interactions that occur in the sample. Therefore to
convert this specimen/standard intensity ratio (known
as the k-ratio) into say weight%, complex mathematical
correction procedures have to be applied which have to
correct the measured X-ray intensities due to the
effects of the electron-specimen interactions. These
can be classified into three categories:
(1) the effect of atomic number (Z)
(2) the effect of absorption (A)
(3) the effect of fluorescence (F)
The procedures used which correct for these effects
are often referred to as ZAF corrections. In this
summary we can only scratch the surface of quantitative
correction procedures (see 7.3) but what we shall do is
explain the phenomena on which these corrections
depend.

## 3.1 THE ATOMIC NUMBER EFFECT
When the electrons are incident on a sample some of
them will penetrate the sample and are scattered, and
others are backscattered out of the sample with little
energy loss. These two processes both affect the
production of X-rays in the sample. The fraction of
electrons that enter the sample and remain there is is
governed by a factor called R. This factor is strongly

dependent on the average atomic number of the specimen. Samples of high atomic number will backscatter more electrons out of the sample and thus have a low value of R, whereas those with low atomic numbers produce less backscattering and subsequently have a high value of R.

It is only the electrons that penetrate the sample that can produce X-rays. The production of X-rays depends on the value of the critical excitation energies, $E_c$, of the specimen and thus depends on its composition. Elements of low atomic number have lower values of $E_c$ and are said to have a greater 'stopping power' per unit mass. The stopping power is referred to as the S factor and is higher for elements of lower atomic number.

The accelerating voltage, $E_o$ of the incident electrons will also affect the values of R and S. A higher accelerating voltage will cause more electrons to be backscattered out of the sample thus reducing R, but the stopping power of the sample, S, may be reduced

Computer programmes have been written which model how the electrons penetrate into a sample taking into account the random nature of the interactions thus determining the shape of the 'teardrop' (fig.4). The effect of accelerating voltage and atomic number on the penetration and ionisation of a sample is shown in figure 8. In general the higher the incident electron energy and the lower the average atomic number of the sample the larger the size of the teardrop.

### 3.2 X-RAY ABSORPTION
When X-rays travel through matter they will be absorbed as they collide and are scattered by the atoms present. If X-rays of intensity $I_o$ are incident on a material of density $\rho$ and thickness $x$, the intensity of the emergent beam, $I_t$, is described by the equation:

$$I_t = I_o e^{\mu \rho x}$$

where $\mu$ is the mass absorption coefficient of the material.

As the X-rays used for microprobe analysis are generated at different depths in the sample, owing to the shape of the teardrop, they are absorbed to different degrees. The greater the distance the X-rays have to travel through the material the more it will be absorbed before it can escape and be detected. The amount of absorption depends therefore on the depth at which the X-ray is generated and the angle at which it leaves the sample on its way to the detector. It is

132

seen (fig.9) that X-rays leaving at low angles will be absorbed more than those leaving at higher angles due to the increased distance the X-ray has to travel before escaping from the sample. The angle between the specimen surface and the detector is called the 'take-off angle' and X-ray detectors are positioned in the microscope so that the 'take-off angle is high so that the absorption of X-rays whilst they are leaving the sample can be minimised. The size of the teardrop also affects the amount of absorption as it determines the distance through the sample that the X-rays have to travel before escaping from the sample.

The mass attenuation coefficient of a material is heavily dependent on atomic number so the composition of the material will affect the amount that an X-ray will be absorbed. In general $\mu$ increases with atomic number so the heavier elements will absorb X-rays more than the lighter elements and also the lower the energy of the X-ray the more likely they are to be absorbed.

For quantitative analysis a correction factor, A, is required which can calculate the amount of absorption a characteristic peak has suffered taking into account
(i)    the shape of the teardrop in the sample
(ii)   the take-off angle and
(iii)  the specimen composition.

### 3.3 X-RAY FLUORESCENCE
X-ray fluorescence, as described in section 2.5, is another way in which X-rays are absorbed whilst leaving the specimen. The effect of X-ray fluorescence on the intensity of characteristic peaks in a specimen is dependent on the accelerating voltage, and the concentration of the exciting and excited elements.

The effect of fluorescence is seen most strongly when the primary X-ray energy just exceeds the critical excitation energy for an X-ray line to be emitted from another element in the sample. This effect is strong when elements in the sample are two atomic numbers apart on the periodic table for elements with $Z > 11$, and one atomic number apart for those with $Z < 11$. For example, in a stainless steel sample containing iron ($Z = 26$) and chromium ($Z = 24$), $FeK\alpha$ X-rays (of energy 6.4keV) can excite the $CrK\alpha$ X-rays (of energy 5.41keV). The effect of this is that the intensity of the $FeK\alpha$ line is reduced and the $CrK\alpha$ line enhanced.

The fluorescence correction (F) in quantitative analysis calculates the effect of fluorescence on the characteristic peaks .

## 4 ED X-RAY INSTRUMENTATION

The components of a typical ED analysis system are shown in figure 10. The system comprises devices to :-
 (a) detect the X-rays
 (b) measure their energy
 (c) display their energy as a spectrum and
 (d) a computer to perform the necessary data processing.
   In the ED system the X-rays pass through a thin beryllium window into the detector which is transparent to X-rays emitted from elements with an atomic number greater than 11 and protects the detector from contamination. The X-ray then produces a charge pulse in the semiconducting crystal in the detector, the magnitude of which is proportional to the energy of the X-ray. This short-lived pulse is converted first into a voltage by the preamplifier and then amplified and converted into a digital signal. This signal then adds one count to the appropriate channel of a multichannel analyser (MCA) which is calibrated to correspond to the energy of the detected electron. It is the content of the MCA which is viewed on a video monitor and forms the ED spectrum.
   We shall now discuss the various elements of the system in more detail to gain a better understanding of how an ED spectrum is obtained.

### 4.1 THE Si(Li) DETECTOR
The most common system used for EDS is the lithium-drifted silicon solid state detector (a schematic representation of the detector is shown in fig.11). When X-rays enter the detector they are absorbed by the silicon atoms and the X-ray photon energy is transferred to photo and Auger electrons. These electrons in turn lose energy by processes which include raising electrons in the semiconductor to the conduction band from the valence band. This process generates a free electron in the conduction band and a positive hole in the valence band which are both free to move when a voltage is applied (fig.12). In the Si(Li) detector the energy required to create a pair is about 3.8eV. The number of electron/hole pairs generated is directly proportional to the energy of the incoming X-ray.
   First we shall talk about the analysing crystal. Ideally the detector would be made out of a pure piece of single crystal silicon but even the purest silicon always contains impurities. These can have the effect

of introducing further energy levels between the valence and conduction bands in the semiconductor, hence the conductivity of the crystal will vary greatly with the amount of impurities. To overcome the problems of impurities lithium is introduced into the crystal by the process of 'drifting'. This process allows the lithium to diffuse into the silicon (at a temperature of 100°C) with an applied electron field controlling the rate of diffusion. Because lithium atoms are very small they can easily move through the crystal lattice and fill holes etc. hence compensating for the native impurities of the crystal. The distribution of lithium in the crystal after drifting is shown in figure 13. In the 'intrinsic' region the lithium concentration is constant and compensates exactly for the crystal impurities. It is only this region in the crystal which is suitable for X-ray detection. The material with a low lithium concentration is removed to expose this material and a thin gold layer, approximately 20 nm thick, is evaporated onto the surface to form a contact. However at the surface of the crystal there is always a layer where neutralisation of the impurities is not complete, leaving excess holes, which is referred to as the dead layer. In this layer the charges generated via X-ray absorption may be trapped in the crystal instead of being detected, thus reducing the size of the charge pulse.

The crystal is operated by connecting a negative voltage of 10-100V to the gold contact at the front of the crystal and obtaining the output signal via another metal electrode connected to the back of the detector.

The Si(Li) detector minimises the baseline current caused by thermal excitation of electrons across the energy gap called the 'leakage current',and is able to withstand the reverse bias voltage at which the crystal is operated. To minimise the unwanted conductivity due to thermal excitation and to ensure that the lithium remains in the intrinsic region, the detector is cooled to liquid nitrogen temperature. This is achieved by connecting the detector ,by means of a copper rod,to a cryostat which is filled constantly with liquid nitrogen. The process of heating up and cooling down the crystal leads to a degradation of its properties.

The limitations on count rate do not originate in the crystal but in the electronics needed to amplify the very small current (about 10-16 coulombs) and convert it into a signal which can easily be measured. The first stage in doing this is the field effect transistor (FET) which acts as a preamplifier. This will now be explained.

## 4.2 THE FET

The problem we are faced with is how to count and quantify the tiny currents produced in the crystal by the incoming X-rays. To do this an FET is used which converts changes in charge, detected at its input (fig.14(a)), to larger changes in voltage at its output (fig.14(b)). The purpose of the preamplifier is to amplify the pulses produced by the X-ray photons with the addition of as little unwanted electronic noise as possible. If we look at the        voltage at the output of the FET (fig.14(b), we see that it goes up in steps as a function of time. The height of these steps are a function of how much charge was collected from the X-ray photon. Therefore the larger the step the greater the energy of the detected photon. To keep the voltage within the linear range of the transistor it is periodically necessary to reset the output to zero and begin again.

At this point it is appropriate to introduce the concept of the 'dead time' associated with the analyser. There are times that X-rays detected by the crystal will not be able to be detected and during that time the detector is said to be 'dead'. Dead time is introduced into the system at various times during the signal processing stage. One of the processes which introduces dead time is when the FET is reset  as just described. The dead time from this source can vary from one acquisition to another. For example, a sample that emits 1000 10keV X-rays per second causes roughly twice as much current to flow through the FET as 1000 5keV X-rays causing the FET to reset itself twice as often thus increasing the dead time.

## 4.3 THE MAIN AMPLIFIER

The next stage in the signal processing is to convert the small step increases produced by the FET (milli-volts) into pulses large enough for pulse height analysis (volts). It is in this amplification process that most of the limitations of the technique are introduced. The resulting output from the amplifier are pulses with a Gaussian or semi-Gaussian shape. Whereas the rise time of the step increases from the FET is very short, the process of amplification has increased the duration of the initial pulses (fig.14(c)). Electronic filters which affect the rise time of the pulses can be characterised by a parameter known as the

136

'time constant'. The larger the time constant the less sensitive the amplifier is to high frequency noise. The time constant however determines the time taken for the pulse to reach its maximum level, given an instantaneous change at the input as produced by the output from the FET. Therefore the magnitude of these time constants determine the time required to process each individual X-ray. This produces a trade-off between the rate at which X-rays can be processed and the accuracy with which each pulse can be measured. To reduce the time taken to process these pulses the time constants are carefully set to control both the rise time and the fall time of the Gaussian shaped pulse. This is called time variant processing.

## 4.4 PULSE PILE-UP REJECTION
Problems arise when a second pulse reaches the main amplifier before an earlier pulse has been analysed. When two such nearly coincidental pulses arrive at the main amplifier the processor cannot distinguish between the two. The result is the two pulses are added together forming a single pulse of enhanced amplitude (fig.14(c)). To prevent this occurring a technique called pulse pile-up rejection is used whereby nearly coincident pulses are rejected. To achieve this a second pulse amplifier is used with a much shorter time constant and hence a faster response time than the main amplifier. This system uses a discriminator which is able to determine the time of arrival of the individual pulses. By measuring the time interval between consecutive pulses and knowing the time constants used to process pulses in the main amplifier the system can determine whether the pulses will overlap. If the discriminator does decide that the pulses will overlap they will be rejected and will not be analysed. However it cannot discriminate between two pulses which arrive simultaneously, which results in an X-ray whose energy is equal to the sum of the two individual energies being recorded. Because of the requirements for fast channel discrimination, pulse pile-up circuits lose their efficiency at low energies when the amplitude of the X-ray events approaches that of the noise level.

Pulse pile-up rejection is another source of dead time in the ED system. Because it is rejecting nearly coincident X-rays, an increase in the rate at which X-rays enter the detector does not necessarily result in an increase in the number of pulses being processed as more pulses will be rejected.

## 4.5 THE MULTICHANNEL PULSE HEIGHT ANALYSER

In order to produce the ED spectrum the magnitude of the output pulses from the main amplifier have to be measured. The pulses are first processed by an analogue-to-digital converter (ADC). This works by allowing the voltage pulse from the main amplifier to charge a capacitor which is then allowed to discharge at a constant rate. The larger the pulse the longer it takes for the capacitor to discharge. The time taken for this to happen is measured and a count is measured to the appropriate channel of a multichannel analyser (MCA). There are typically 1024 channels in the MCA which are often calibrated so that each channel is 10eV thus giving a whole spectrum range of 10.2keV. When using higher accelerating voltages, as in the TEM, it is often calibrated at 20eV per channel so the higher energy X-ray lines can be detected. It is the contents of the MCA which form the ED spectrum and can be viewed and manipulated on on a visual display unit. A close look at a typical spectrum (fig.15) the individual channels calibrated to correspond to the energy of the X-rays detected.

## 5 THE ELECTRON COLUMN

ED spectrometers are fitted on a wide range of opto-electronic instruments so the ED user should appreciate the similarities and differences between the different instruments and their particular application to X-ray microanalysis. Modern electron columns can be roughly classified into three categories:-
   (1) scanning electron microscopes (SEMs)
   (2) electron microprobe
   (3) transmission electron microscopes (TEMs)

## 5.1 SCANNING ELECTRON MICROSCOPES

This is the most common type of electron column used and is designed primarily to provide images of high spatial resolution using secondary electron images. The accelerating voltage is typically 5-30KV. Because of the sensitivity of the secondary electron image to topography and the good depth of field of the imaging system, the SEM is used routinely to look at rough samples such as fracture surfaces and three-dimensional objects. The specimens are held in a tiltable stage so that different viewing angles can be used. To obtain the high magnification in the SEM ( up to about 50 000 times) requires that the incident beam diameter is very

small, generally less than 100A. This means that the
beam currents used are very small thus producing a low-
energy flux of emitted X-rays. As the ED spectrometer
has a better detection efficiency than the WD
spectrometer and it is not limited by geometry it is
the most popular system to be fitted to the SEM. The
detector can be positioned close to the specimen so
collecting a larger solid angle of X-rays. Some SEMs
may be fitted with an ED spectrometer and WD spectrometer. The WD spectrometer may be fitted with an
analysing crystal dedicated to the analysis of light
elements or may be used to overcome the problems of
peak overlaps in the ED system when looking for
elements present in low concentrations.

## 5.2 THE ELECTRON MICROPROBE
In many aspects the electron microprobe is similar to
the SEM but it differs from the SEM due to its intended
use. Whereas the SEM is designed with the emphasis on
high resolution imaging, the electron microprobe is
designed with analysis as its main objective.
Essentially the electron microprobe is designed to
deliver stable beam currents of high intensity and
usually has scanning and imaging facilities. Most
microprobes are fitted with wavelength dispersive
spectrometers which require a much higher beam current
to provide sufficient characteristic X-rays to make WD
analysis practical. The emphasis on quantitative X-ray
analysis against elemental standards in the microprobe
requires the use of flat specimens mounted on a fixed
geometry so that the necessary corrections can be made.
The maximum resolution is lower on the microprobe than
the SEM due to the larger incident beam.
　　The difference in primary function of the SEM and
electron microprobe is usually accompanied by different
outlooks on the part of the operator. The complex
operation associated with the electron microprobe with
WD spectrometers necessitates that the operator is
specialised in the field of X-ray analysis.

## 5.3 THE TRANSMISSION ELECTRON MICROSCOPE
The third type of electron column is the transmission
electron microscope. In the TEM the sample has to be
thin enough to allow the high-energy electrons to pass
through it. The image is formed by subjecting the
sample to a widely dispersed and homogeneous flux of
electrons, instead of a focused beam, and the
electrons which are able to pass through the sample are

focused down onto a luminescent plate below the specimen. As the samples are very thin the amount of scattering of electrons into the sample is much reduced compared to that in a bulk sample giving the TEM a much greater resolution than the SEM. This also means that the resolution for X-ray analysis is greater as we are only concerned with the X-rays which emanate from the top of the 'teardrop' as shown in figure 4.

The modern variation of the TEM is the scanning transmission electron microscope (STEM) which combines the principles of the SEM and TEM. In this case a finely focused beam of electrons (of approximately 1nm diameter) is scanned over the electron transparent specimen, and the image is produced by detecting the electrons that are transmitted through the specimen. The addition of scanning facilities to the TEM enables the operator to select particular points for X-ray analysis and produce X-ray maps of the specimen with a very high resolution.

Crystal spectrometers are very rarely incorporated in the TEM, as the X-ray intensities emitted from the thin samples are very low; so it is the ED spectrometer which is most commonly found. The principles of analysis are the same as with the SEM, but a particular difficulty with thin specimens is that the X-ray intensity is dependent on the thickness of the sample which can vary from point to point and cannot be directly measured. (Special techniques for quantitative analysis have been developed which can account for this.)

## 6 SPECTRUM ACQUISITION

As we have seen (fig.15) an energy dispersive spectrum is displayed as a histogram with the x-axis labelled in energy units (eV) and the y-axis in X-ray counts. When a spectrum is acquired all the elements in the spectrum appear simultaneously. The time taken for the characteristic peak from a particular element in the sample to appear depends mainly on its abundance. For bulk samples (elements in quantities of greater than 10%) it takes approximately 10 seconds for the characteristic peaks to appear. For trace elements at much lower concentrations a longer acquisition time is required as the counting statistics for the analysis become more important. Typically a counting time of about 100 seconds is used which is often increased when looking at thin samples which produce much fewer X-rays.

## 6.1 DEAD TIME

When considering the time taken to acquire a spectra one of the first things to consider is the dead time (section 4.2). When the system is 'dead' no other pulses entering the system will be processed. The amount of dead time produced is proportional to the rate at which X-rays enter the detector. In figure 16 the variation in output count rate is plotted against dead time. It is seen that after a certain point an increase in input count rate results in a decrease in the output count rate owing to the large amount of dead time being produced in the system. If we take points A and B in fig.16 we can see that the output count rate is the same but because of the large amount of dead time associated with B it would take much longer to acquire a spectrum at this point than it would using the input rate at A although the spectra would be the same. For effective operation of the ED system the dead time should be kept below about 30%.

   This introduces a concept called 'live time' which can be thought of as the actual number of seconds that the processor is counting when acquiring a spectrum. For example, if the normal time (or clock time) taken to acquire a spectrum is 100 seconds, then if the detector was operating with a dead time of 25% then the processor was only counting for 75 seconds of that time. If we set the 'live time' to be 100 seconds with the same dead time the time for the spectrum to be acquired would be 100 seconds counting time plus 25 seconds dead time which is 125 seconds. This means that the time taken to acquire the spectrum will vary with count rate but the time spent actually measuring will always be 100 seconds.

## 6.2 SPECTRAL ARTEFACTS

In the ED spectrum a certain number of artefacts occur which are produced by the detector and the pulse processor. This means that peaks can appear in the ED spectrum which do not correspond to the characteristic X-ray peaks but are produced via other mechanisms.

**Silicon escape peaks:** The process of measuring the energy of an X-ray relies on the fact that when the X-ray enters the detector all its energy is used to produce electron/hole pairs in the silicon crystal.

There is a chance however that an incoming X-ray can use some of its energy to produce a silicon X-ray which may either escape from the detector or be reabsorbed producing more electron/hole pairs. If this occurs the energy of the silicon X-ray peak (1.74KeV) will be lost from the measurement of the original X-ray. This produces an artificial peak, called an escape peak, whose energy is 1.74KeV less than the parent peak (fig.17). It is possible that both the SiK$\alpha$ and SiK$\beta$ X-rays can be generated but as the probability of exciting the K$\beta$ peak is only 2% of the K$\alpha$ only one escape peak is usually seen.

Silicon escape peaks will only appear when the incoming X-rays have an energy greater than the critical excitation energy for silicon (1.83KeV). The escape peaks are only obvious when associated with high intensity X-ray lines.

The silicon X-rays generated via this process can themselves be absorbed by the crystal and consequently detected. Thus a small silicon peak can be observed in the spectrum which has not originated in the sample but in the detector.

**Sum peaks:** As already mentioned (4.4) the processor is unable to distinguish between two X-rays that arrive simultaneously at the detector. When this occurs an X-ray is measured with an energy equal to the sum of the two individual X-ray energies. This produces an X-ray peak at a higher energy value in the spectrum (fig.17). Sum peaks are only observed associated with major peaks in the spectrum at values equal to twice the energy of the parent peak. In figure 17 the sum peak of the TiK$\alpha$ peak whose energy is 4.5KeV appears at 9KeV.

**Absorption edges:** Absorption edges appear in the spectrum as a sudden drop in the level of the background radiation at a point slightly above the energy of the SiK$\alpha$ X-ray. This is because these X-rays can be preferentially absorbed by the silicon in the dead layer of the detector. The magnitude of this drop is therefore related to the dead-layer thickness and can be used to monitor it. If the dead layer is too large crystals will be rejected during manufacture.

**Peak distortion:** The main contribution to peak distortion is due to incomplete charge collection in the detector. At the faces and edges of the detector the electron/hole pairs produced by the incoming radiation have a greater probability of recombining rather than being pulled apart by the reverse bias. If this happens then the subsequent measurement of the X-ray will be lower than it should be. The effect is that the Gaussian peak is distorted at the low-energy side which is known as tailing.

**Stray radiation:** When analysing X-ray spectra it is possible that stray radiation occurs which has originated from places other than the sample. Figure 18 illustrates where such stray radiation could originate from in the SEM. The problem is a lot worse for TEMs as the electrons have much higher energies and a much smaller specimen stage.

## 7 ANALYSIS

Having dispensed with the theory of X-ray generation and detection we now at last turn our attention to analysis. The job in hand for the ED user is to acquire a 'raw' X-ray spectrum and process it to provide 'meaningful' results of the analysed specimen. In modern ED systems most aspects of analysis are automated, so at the push of a button answers can be obtained. This makes the ED system a relatively simple instrument to use. However life is never that simple as the user is often required to make informed choices during the processing and interpretation of X-ray spectra. It will be revealed as we start discussing the practical aspects of analysis that a fundamental understanding of the theory as outlined in the previous sections is important for the interpretation of X-ray spectra.

### 7.1 QUALITATIVE ANALYSIS
The simplest form of operation for the ED system is qualitative analysis which implies that the results will tell us what is there, but not necessarily how much. To aid the ED user to identify the characteristic peaks all modern ED systems have 'look up' tables which contain the energy positions of the major characteristic peaks of all the elements. By identifying the energy position of an unknown peak, the computer will tell you the possible options for that energy. As there may be many overlaps of the characteristic peaks in the ED spectrum it is at this point that the user has to use his judgement in

matching the peaks with their theoretical positions. Once identified the peaks can be labelled with their particular elements.

For the correct identification of an element it is essential that the system is calibrated as the centres of the characteristic peaks must fall within one channel of the theoretical energy value. Most systems have automatic calibration routines whereby the system is calibrated by acquiring a spectrum from a known sample. The calibration should be checked regularly as the system can 'drift' out of calibration in two ways:

(i)   zero drift - whereby the whole spectrum moves in energy value and
(ii)  gain drift - where the relative positions of the characteristic peaks changes.

The process of calibration only takes a few minutes but if the system is maintained at a constant room temperature it should not vary significantly over a period of a few months.

When trying to determine 'what is there' we have to consider the minimum detection limits of the system so we have to bear in mind that other elements may be present in quantities lower than the detection limit.

## 7.1.1 Peak overlaps

The presence of overlapping peaks gives rise to the greatest source of error, or uncertainty, in qualitative analysis. The problem is best explained by example.

A notorious overlap is that of the SK$\alpha$ line with the MoM$\alpha$ line. This is a common problem for the metallurgist when looking for sulphidation in high-strength steels which often contain molybdenum. Figure 19 contains a secondary electron image of a corroded stainless steel showing two points which were analysed. The spectra obtained showed that either molybdenum or sulphur were present in each. The best way to see which is present is to use a sufficiently high accelerating voltage (in this case 30kV) so that the high energy MoK$\alpha$ line is excited.

From the two spectra we can see that point 1 contains molybdenum and is probably a carbide and point 2 contains no molybdenum so therefore contains sulphur.

## 7.1.2 Light element analysis

So far we have only concerned ourselves with the analysis of elements with atomic numbers greater than 11 (sodium). Conventional detectors fitted with a beryllium window cannot detect the X-rays emitted from the light elements as they are absorbed in the window before reaching the detector. To detect these X-rays this window has to be removed or replaced with a thinner window which is transparent to the X-rays. This enables elements down to boron to be visible to the detector but the intensities of the peaks are very low and the resolution is poor so they are only used mainly for qualitative analysis.

Figure 20 shows the ED spectra obtained from a piece of pure silica ($SiO_2$) with and without the beryllium window.


## 7.1.3 X-ray mapping

A simple method for displaying the relative distribution of a single element in a sample is that of X-ray mapping. To produce an X-ray map an area of interest on the sample is first identified via normal imaging methods. When the electron beam is scanned over the surface the intensity of a selected X-ray line is used to modulate the brightness of the image on the SEM video display. To select a particular X-ray line a 'window' is placed over the peak in the spectrum, which determines the energy range of the X-rays that will go to form the X-ray map. The area covered by the window includes both characteristic and background radiation. As it is only the characteristic X-rays that we are interested in, a convenient feature on many ED systems is the ability to remove the background counts so they do not add to the X-ray map. This is done by mapping X-rays entering the detector only when the peak intensity exceeds the background level.

Figure 21 contains a secondary electron image of inclusions found in a high temperature steel (which were probably introduced during casting) along with X-ray maps for aluminium and oxygen. The increased density of dots associated with the inclusions seen in the X-ray maps show that they are rich in these elements. Point analyses had revealed that the inclusions only contained these elements but the X-ray maps give a simple visible representation of the elemental distribution which is easily compared to the SE image.

Another example of X-ray mapping is shown in figure 22 where a number of X-ray images were taken of an oxidised silicon nitride ceramic. The elemental distribution in the different phases in the sample is easily seen. This example was taken from the video display of a modern ED system where the grey scales associated with the X-ray maps can be given colour scales which can add further impact to the reporting of these results.

In the case of elements present in low concentrations or those with low count rates (such as oxygen and nitrogen in the examples above) it requires a much longer time to acquire a well defined X-ray map and an X-ray map taken with two few X-ray counts will lack detail. The use of frame stores and image averaging systems enable X-ray images to be acquired from many scans across the sample until an image of sufficient quality is obtained.

### 7.1.4 Line profiles
Another method of presenting the results from a qualitative analysis is line profiles. This technique shows how the X-ray intensity of a particular element varies along a line across the sample. Figure 23 shows a backscattered electron image of a two phase titanium alloy. The two line profiles for molybdenum and aluminium reflect the change in X-ray intensity for these elements when the electron beam was scanned across the line shown on the image.

### 7.1.5 Point analyses
In most investigations the microscopist will observe a particular feature in the electron image of which an analysis is desired, be it either qualitative or quantitative. In this case the electron beam, in either the SEM, TEM or STEM, is positioned on the feature of interest and remains stationary whilst a spectrum is acquired. The size of the analysed volume will depend on the sample and the analysis conditions.

Figure 24 contains a TEM image of a high-temperature titanium alloy and the spectrum obtained from a point analysis of the precipitate as marked. It is shown that it was a silicide rich in zirconium and titanium. Note that the copper peak which appears in the spectrum has originated from the microscope. This is a common problem with older machines not originally designed for analytical microscopy.

### 7.1.6 Image processing and ED analysis
Modern ED systems are now able to combine the process of image analysis and chemical analysis. Images can be stored digitally (e.g. secondary or backscattered electron images) and then the image can be analysed to

produce a variety of measurements. For example from the grey scale in the image features can be identified, sized and then the system can automatically send the electron beam to the stored location of the feature for chemical analysis. From processing the spectra the features could be classed by size and chemical composition and other calculations such as area fractions could be carried out.

An example of the combined use of image and X-ray analysis is shown in fig.25. Here the intermetallic phases observed in the silicon/aluminium alloy have been identified, sized and analysed. Sophisticated software packages enable the user to display the results in a variety of ways. In this case they are presented in the form of a three-dimensional histogram indicating the number of features measured, their size, and chemical classification. The use of colour graphics again gives impact to these results which enables the user to provide a professional presentation.

## 7.2 OPERATING CONDITIONS
To obtain the best results with an ED system it is important to use the correct instrument operating conditions.

**Accelerating voltage:** This is the most important operating condition to consider. The energy of the incident electron beam has to be sufficient to excite the characteristic X-rays of the elements present in the sample as it is easy to 'miss' an element in the sample purely because you have not excited its X-rays. As well as this there are other factors to consider. Because the accelerating voltage affects the size of the teardrop in the sample, the larger the incident electron energy the larger the analysis volume. To obtain the best spatial resolution the lowest accelerating voltage which will give adequate X-ray emission should be used. This also minimises the amount of absorption that the X-rays suffer as they leave the sample, because they are generated nearer to the surface.

**Beam current:** The intensity of the X-rays emitted from a sample is directly proportional to the current of the exciting electron beam. It is this beam current that has to be adjusted when setting the dead time, as it is required that sufficient X-rays are generated so that it does not take too long to acquire a spectrum but the X-ray intensity should not be too large to incorporate excessive dead time.

The beam current and hence the X-ray emission will increase with accelerating voltage. However a large beam current will increase the analysis volume thus degrading the resolution of both imaging and analysis. The choice of beam current is therefore a compromise between obtaining good spatial resolution and sufficient X-ray intensity for analysis.

In many cases the operator is required to analyse samples on a regular basis, such as production control of heat treated or coated specimens. From day to day the conditions in the microscope will change due to room temperature, filament wear, and gradual contamination of the electron column. To ensure that analysis conditions are as near identical as possible it is advantageous to monitor the beam current. Indeed, in the analysis of specimens where long acquisition times are involved the beam current should be monitored regularly during the analysis. There are two main ways of doing this; one is to monitor the absorbed current through the sample, the other is to use a Faraday cup which can be inserted into the column. The latter is more efficient and should the beam current drift it can be normalised to compensate.

**Specimen geometry:** In the electron microprobe the specimen holder and X-ray optics are designed so that the specimen surface is perpendicular to the electron beam. This is because the quantitative correction procedures rely on this geometry to calculate the effect of absorption of X-rays as they leave the sample. The SEM however is frequently used to examine rough samples so the beam may strike the sample at some other angle. The sample tilt affects the average depth of the interaction volume (see fig.26). The greater the angle of tilt away from the normal angle of incidence the closer the volume is to the surface. Tilting the sample can be used to estimate the thickness of surface films or to try and confine the analysis to the surface of the specimen.

## 7.3 SAMPLES

### 7.3.1 Bulk specimens for the SEM
As the SEM is especially good at examining rough surfaces it is such specimens that are frequently

analysed using EDS. Scanning the beam over rough samples does not normally lend itself to quantitative analysis. Because the ZAF corrections used for quantitative analysis require that the take-off angle of the X-rays leaving the sample is known. It is important therefore when using ZAF that the samples are flat and polished.

One has to be careful when analysing rough samples, even qualitatively, as at first glance what appears in the spectrum will not necessarily have originated in the probe volume. Figure 27 shows the effect of surface roughness on the collection of X-rays. It is seen that one cannot analyse a hole as the X-rays cannot escape to the detector and stray X-rays can be produced by backscattered electrons being incident on peaks in the sample surface.

When examining non-conducting samples such as ceramics problems arise as the specimen will charge up under the electron beam. This will result in the beam being deflected by the sample thus distorting the image and affecting the X-ray intensity emitted. This is because the incident electrons are unable to escape so a method is required to allow these electrons to flow to earth. To do this the specimen may be coated, by vacuum evaporation, with a conducting material such as carbon or silver.

A problem that most ED operators come across is that many conducting specimens are submitted for analysis which are mounted in non-conducting resin or bakelite which again creates the problem of specimen charging. This can be overcome by coating the sample or by making a connection using conducting paint from the specimen to the specimen stage. An easier option, however, is to mount the specimens in conducting bakelite to start with!

### 7.3.2 Thin specimens for the TEM
Specimens for the TEM come in two forms, thin foils and extraction replicas.

**Thin foils:** Thin metal foils usually take the form of a disc, 3mm in diameter, which has been thinned so a hole has formed by the centre. The region next to the hole is usually thin enough so that electrons can pass through it to form the image in the TEM. The thickness of the specimen however will vary, which causes problems for quantitative analysis and the comparison of spectra because the intensity of X-rays will vary with the thickness.

**Extraction replicas:** This is a useful technique to isolate individual phases within the microstructure. Problems occur, however, with quantitative analysis due to variations in particle size and shape. There are ways of correcting for these effects which are being incorporated in the latest correction software. A further problem is that the support grid contributes to the spectrum. Copper, aluminium or beryllium grids can be used depending on sample composition.

## 7.4 SEMI-QUANTITATIVE ANALYSIS
After establishing which elements are present in a sample the next step is to say how much of each element is present. A first approximation is known as semi-quantitative analysis. One method used to provide quick semi-quantitative estimates of elemental concentrations uses the ratio of elemental intensity of the unknown element in the sample to the intensity from the pure element (see sect.3). The latter can be obtained by:
(i)  a measurement, at the same operating conditions, on a pure element standard or a compound standard of known composition or
(2)  a calculated value knowing the accelerating voltage, beam current, beryllium window thickness and detector efficiency.
This specimen standard intensity is known as the k-ratio. This simple ratio corrects for the different excitation efficiencies of different elements but ignores the effects of the other elements in the sample. Nevertheless the results are often adequate to answer the analytical questions being asked and are most useful for comparison of spectra when a fully quantitative approach can often be more than is needed to solve the problem. If standards are obtained with compositions as similar to the unknown as possible semi-quantitative analysis can provide a reasonable quantitative result.

For example, if one is trying to establish the aluminium concentration in aluminised coatings, by obtaining a sample in which the aluminium concentration in the sample is known one can use semi-quantitative analysis to compare the aluminium content in the unknown coatings with reference to the standard.

In order to determine the X-ray counts within an X-ray peak which are contributed by the characteristic X-rays of the appropriate element the following processes must be carried out:
(1)  peak identification and choice of X-ray line for

150

analysis
(2) determination of the background radiation and subtraction from the spectrum
(3) removal of the spectral artefacts
(4) deconvolution of overlapping peaks

**Choice of X-ray line:** When an element produces more than one spectral line only one of them is required for analysis. For elements of atomic number < 30 there is little choice as only the K lines fall in the detectable energy range. For heavier elements, however, it is usually necessary to choose either the L or even the M line as the energy of the K lines are too high. When choosing which line to use for a particular element, first select the line which has the greatest intensity. If this line overlaps with another spectral line it may be appropriate to chose the next most intense line as this simplifies the processing of the spectra. It is not always possible to find a peak which does not overlap with another so corrections are made which attempt to separate interfering peaks which will be explained shortly.

**Background removal:** There are a number of methods in which the background level can be determined and subsequently removed from the spectrum.

One method requires the operator to select a number of points on the spectrum which are devoid of characteristic peaks and the analyser computes a curve to the assigned points which should correspond to the background level, or the computer may calculate its own background fit knowing certain analysis conditions such as accelerating voltage, take-off angle etc.

The most popular method though is the use of a digital filter (referred to as a 'top hat' filter) which is able to distinguish between the slope in the spectrum where only background is present and the slope in the vicinity of a characteristic peak. It acts as a frequency filter, filtering out the low frequency background leaving the higher frequency characteristic peaks.

**Deconvolution of overlapping peaks:** Once the background has been removed and the spectral artefacts corrected for, the next stop is to evaluate the intensities of the X-ray peaks. This is straightforward if the peaks do not overlap as the number of X-ray counts is equal to the area under the peak (i.e. the peak is integrated to give the number of counts). If peaks do overlap this job is made more difficult as

151

the overlapping peaks have to be separated before the number of counts attributed by the different elements can be calculated and the technique used to do this is called deconvolution or peak stripping.

One method is that the computer synthesises the shape of the major interfering peak by acquiring the same peak on a pure element then normalising it using a least squares fit routine to match the peak from the specimen. This can then be subtracted from the spectrum leaving the minor overlapping peak behind.

**Trace elements and minimum detection levels:** One question of concern for the potential ED user is the sensitivity of the technique to small elemental concentrations. The minimum detection level of an element in the ED spectrum is determined by the amount a peak has to rise above the average background level before it can be detected. Here we have to consider the statistics of the counting process. The smallest detectable peak is defined as three standard deviations (s.d.) greater than the background counts where the s.d. is equal to the square root of the number of counts.

**Example:** In figure 28(a) a spectrum from a sample containing cobalt, nickel, chromium and tungsten is displayed showing the computed profile of the background radiation. In figure 28(b) the background level has been subtracted leaving just the characteristic peaks. The computer is now able to calculate the areas under the selected peaks for each element which are those outlined in black on the spectrum and the relative intensities are printed out. The semi-quantitative results can now be calculated.

## 7.5 QUANTITATIVE ANALYSIS

Quantitative analysis takes the process of semi-quantitative analysis a step further by calculating the actual concentrations of the elements in the sample, in either weight% or atomic%. It is based upon the ZAF correction procedures which correct for the effects of atomic number (Z), absorption (A), and fluorescence (F) as described in section 3. Quantitative analysis must proceed through the stages as semi-quantitative analysis but finishes with a ZAF correction.

It is in undertaking quantitative analysis where the casual users of ED systems are likely to get into trouble, and to have blind faith in the computed results without having a basic understanding of how

they are obtained can lead to misinterpretation when
answering questions on accuracy while reporting the
results. Most ED manufacturers quote accuracies of
about 0.1 weight% for the ZAF calculation of elemental
concentrations for elements greater than atomic number
11. This kind of accuracy can only be achieved
provided you have the ideal sample (e.g. flat,
conducting, homogenous). Many systems quote the results
to an accuracy of three decimal places and a common
mistake is to report this accuracy as being correct!

### 7.5.1 Quantitative corrections – bulk samples
Having acquired the k-ratios by semi-quant (7.4) these
estimated concentrations are then used to calculate the
magnitude of the Z,A and F factors that have to be
applied which are dependent on the composition of the
sample. Once calculated these factors are applied to
the original k-ratios to yield a second estimated value
of elemental concentrations. This improved value is
then used to recalculate the ZAF factors which are then
applied to the new estimate and so on. This iterative
process will eventually converge within about 3 or 4
loops to give a value which will be as near as possible
to the real elemental composition.

An important point to note is that the ZAF
calculations make a number of assumptions before
proceeding through the analysis. They assume the
sample is microscopically flat and homogenous within
the analysis volume as they rely on these facts for
their calculations.

Figure 29 compares the results from a fully
quantitative analysis, on the same spectrum used for
the semi-quantitative analysis in figure 27. The
corrected results show the change in values of the
elemental concentrations and the amount of correction
applied.

**Standardless analysis:** Quantitative analysis in the ED
system can also be achieved without standards and there
are a number of commercially available techniques which
do this. One method uses virtual standards and here
spectra are acquired from standards for all the
elements when the system is first commissioned. They
are stored in the computer memory and are then
used for the determination of k-ratios as described.
For example the spectra from a pure iron standard
obtained using accelerating voltages of 10, 20, and
30kV would be stored to provide a reference for
analysis. When using standardless analysis the

operating conditions, such as operating voltage and working distance, must be the same as the conditions used to acquire the reference data. The ZAF procedures can use these calculated k-ratios and proceed the same as if standards are used.

### 7.3.2 Quantitative corrections – thin samples
The quantitative analysis of thin foil samples should be much simpler since the major correction (absorption) can also be ignored along with fluorescence as there is no teardrop (fig.4). Unfortunately the thickness of such samples tends to vary locally from point to point. This affects the X-ray intensities as the thicker the sample the greater the chance an atom will be ionised and thus generate X-rays. There are various empirical methods of calculating the thickness, such as measuring carbon build up but these are beyond the scope of this chapter.

## 8 GENERAL COMMENTS

Most of the recent developments in EDS are in computer software. Skilled and dedicated users of ED are continually pushing these developments and there are many technical papers detailing the latest improvements. For example, quantitative analysis of ultra light elements (using a thin window) is now achievable, but generally these esoteric analyses are done in circumstances far removed from an industrial environment.

When using EDS, rather than attempt to stretch the technique to its limits, it is arguably better to work in an area where results are more certain than uncertain. There is no point in undertaking fully quantitative analysis where semi-quantitative analysis will do. If the problem can be resolved by comparing known spectra with the unknown (and many of the problems encountered in materials science are of this nature), then this is an area where ED will give a good solid result and such circumstances should be exploited. When studying the results from quantitative analyses one has to be extremely cautious, as variations of less than 1 weight%, say, seen between different analyses carried out on the same sample may be attributed to many factors, such as surface roughness, other than an actual change in composition of the sample. If such small changes in concentration are of interest then EDS on the SEM is not necessarily the best technique to use and WDS, if available, is better.

Anyone who has read this far will have realised that the technique, though simple to operate, is complicated. The reason for this monograph was to acquaint new scientists with EDS, and if it serves to stimulate more people to become more dedicated to EDS then the information contained therein will not be wasted. The technique has a wide range of applications and there is no limit to the number of opportunities that can be exploited. It is hoped that new scientists, having studied this paper, can appreciate that there can be a dedicated career in X-ray analysis and it is not just a simple technique for the casual user.

## 9 ACKNOWLEDGEMENTS

I would like to thank my colleagues in the Department of Wrought Technology and Process Physics at Rolls-Royce who have given me a lot of support in compiling this chapter. I would also like to thank Link Analytical Ltd. for their cooperation in providing some of the work reported.

## 10 GENERAL REFERENCES

1.  V D Scott and G Love: Quantitative Electron-Probe Microanalysis. Chichester, Ellis Horwood Ltd.(1983)

2.  J A Chandler: X-ray Microanalysis in the Electron Microscope.(Ed. A M Glauert) North Holland Publishing Company (1977)

3.  S J B Reed: Electron Microprobe Analysis. Cambridge University Press (1975)

4.  Ed. D Vaughan: Energy-Dispersive X-ray Microanalysis - An Introduction. Kevex Corporation (1983).

5.  Ed. F Maurice, L Merry, R Tixier: Microanalysis and Scanning Electron Microscopy. ORSAY Les Editions de Physique (1980).

6.  J C Russ: Principles of EDAX Analysis on the Electron MIcroscope. Laboratory Workshop Experiments and Experiments Lecture Notes. EDAX International Inc.

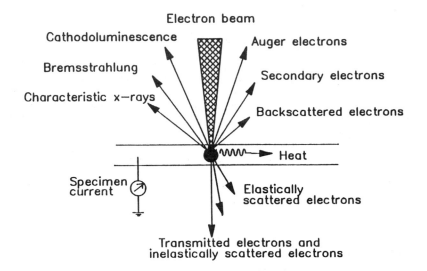

Figure 1. The principal effects from the interaction of
an electron beam with a solid specimen

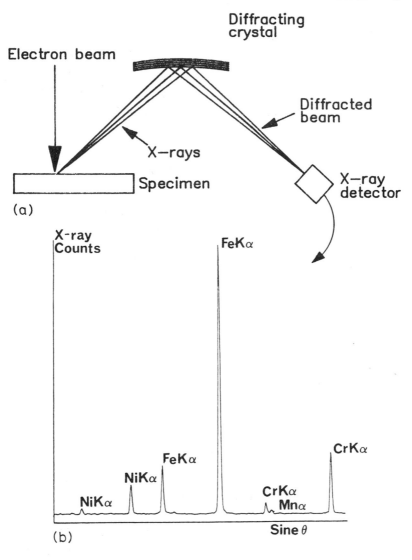

Figure 2. (a) Detection of X-rays by a wavelength
dispersive X-ray spectrometer
(b) Spectrum obtained from a stainless steel
sample via WDS (accelerating voltage
15kV)

157

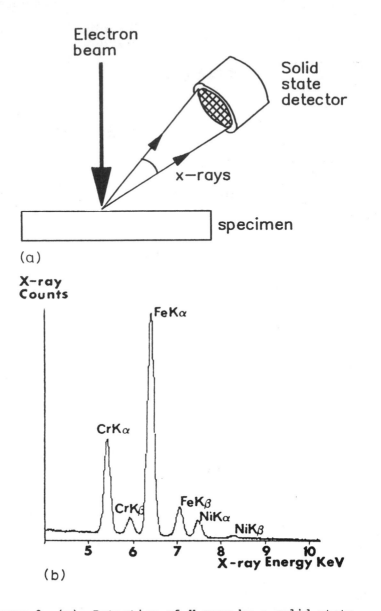

Figure 3. (a) Detection of X-rays by a solid state
energy dispersive X-ray spectrometer.
(b) Spectrum obtained from a stainless steel
sample via EDS (accelerating voltage
15kV)

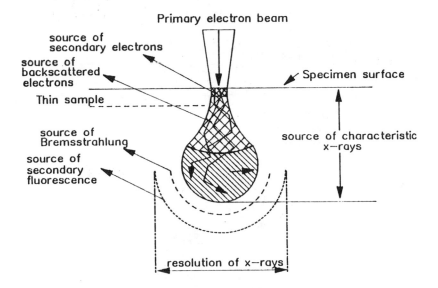

Figure 4. Illustration of the interaction volumes for various electron-specimen interactions

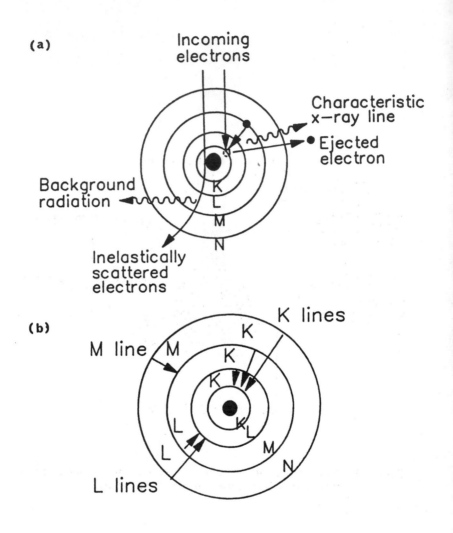

Figure 5. Classical models showing:
(a)    the production of characteristic X-rays
       and background radiation
(b)    some typical line types observed in
       X-ray spectra

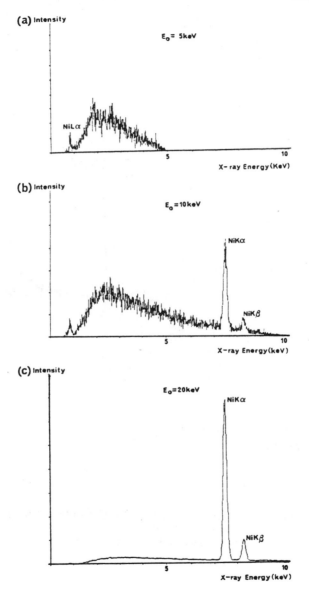

Figure 6. X-ray spectra obtained from a pure nickel
sample using an accelerating voltages of
(a) 5kV, (b) 10kV and (c) 20kV

(a)

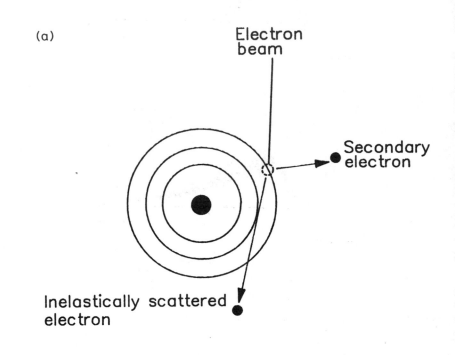

Electron beam

Secondary electron

Inelastically scattered electron

Figure 7. Schematic representations showing the
production of:
(a)   secondary electrons

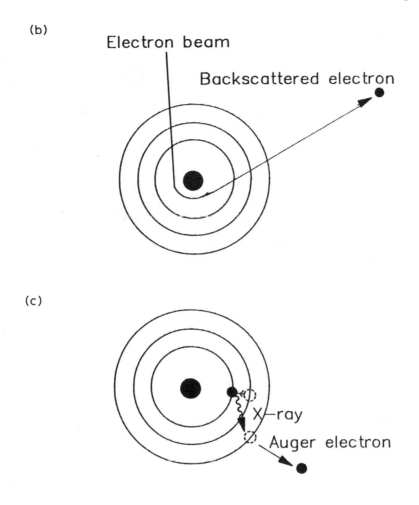

Figure 7. Schematic representations showing the
production of:
(b)  backscattered electrons
(c)  Auger electrons

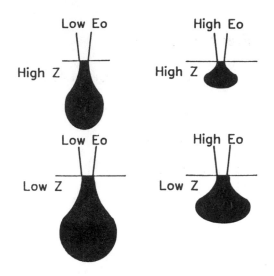

Figure 8. The effect of accelerating voltage and atomic number on the X-ray analysis volume

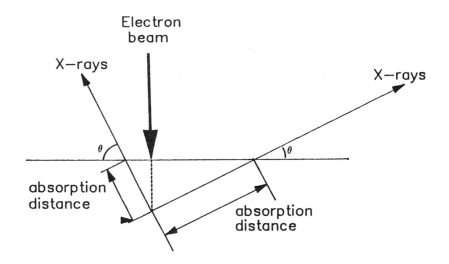

Figure 9. Illustration of how absorption in a thick specimen is affected by the angle at which the X-rays leave the specimen

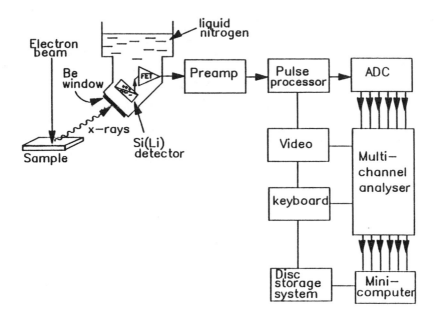

Figure 10.The components of a typical ED system

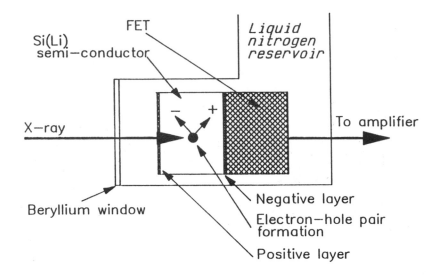

Figure 11.Schematic representation of the Si(Li) solid
state X-ray detector

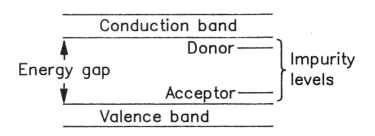

Figure 12.Representation of the energy bands in a
semiconductor

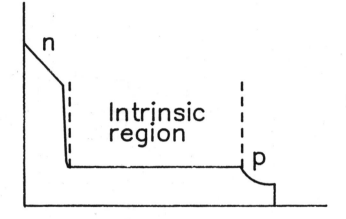

Figure 13.The distribution of lithium in the silicon
crystal after diffusion

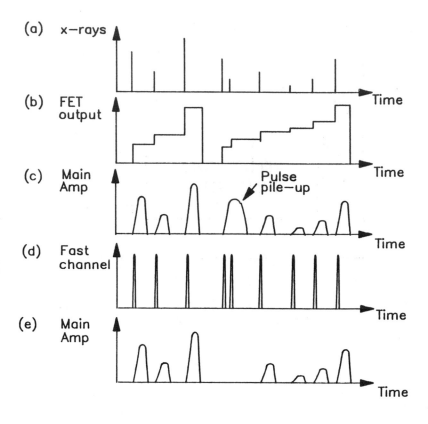

Figure 14. Illustration of pulse processing by the
different electronic devices in the ED
system

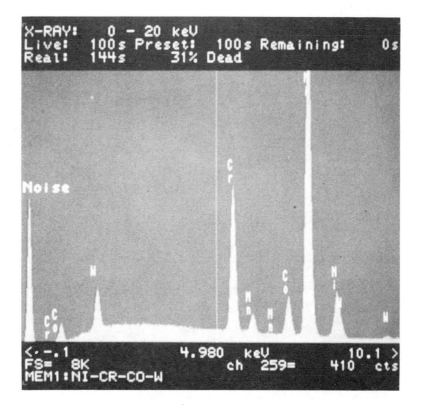

Figure 15.A typical ED spectrum showing that it
comprises a series of individual channels

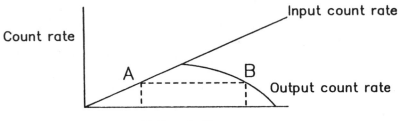

Figure 16. The variation of output count rate with dead time

170

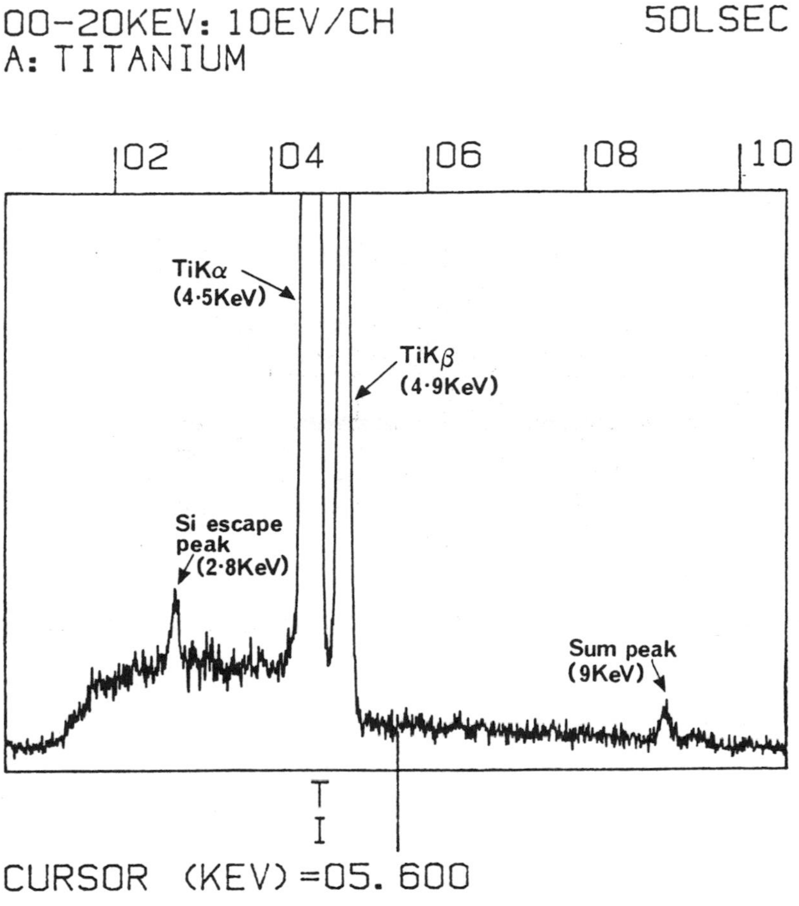

Figure 17. Spectrum obtained from a pure titanium sample showing examples of spectrum artefacts

171

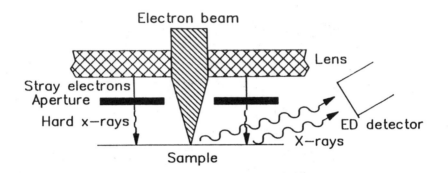

Figure 18.Examples of where stray radiation could
originate in the SEM

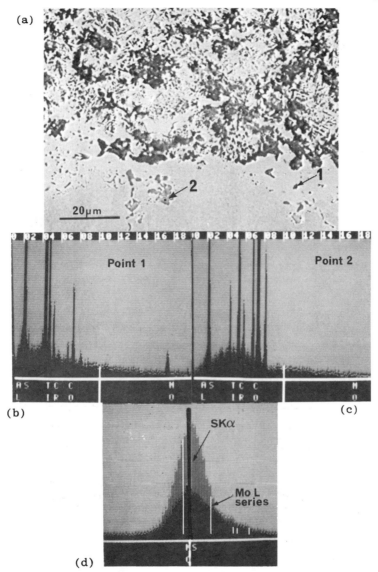

Figure 19.(a)   Secondary electron image of a corroded
                stainless steel sample
        (b)   Spectrum obtained from point 1
        (c)   Spectrum obtained from point 2
        (d)   Comparison of the overlapping SKα and
                MoL lines

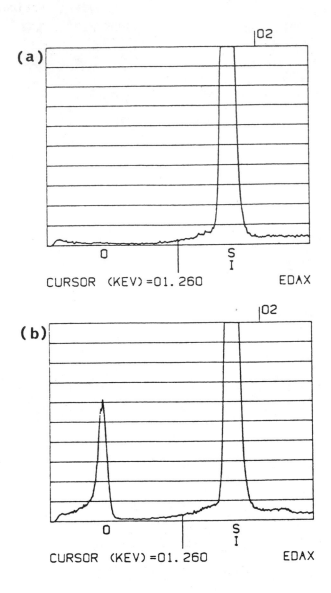

Figure 20.ED spectra obtained from pure $SiO_2$:
   (a)   with the Be window of the detector in
          place
   (b)   without the window

174

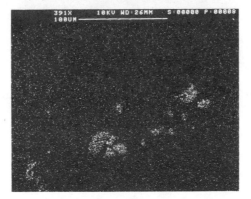

Figure 21.(a)  Secondary electron image of inclusions
found in a high-temperature steel
(b)  X-ray map for aluminium(using AlKα line)
(c)  X-ray map for oxygen (using OKα line)

175

Figure 22.Example of X-ray mapping of an oxidised
silicon nitride ceramic
(Courtesy Link Analytical Ltd)

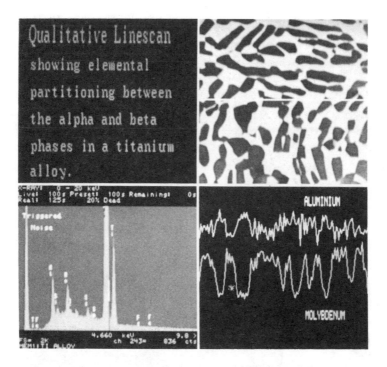

Figure 23. Example of line profiles
(Courtesy Link Analytical Ltd)

Figure 24. Example of the combined use of image analysis
and chemical analysis in the ED system
(Courtesy Link Analytical Ltd)

178

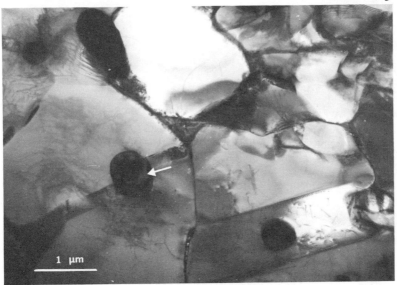

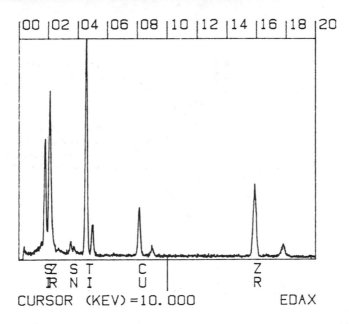

Figure 25 (a) TEM micrograph of a high-temperature titanium alloy and (b) the ED spectrum obtained from a point analysis of the precipitate indicated (Courtesy S McKenzie and R Bolingbroke, Rolls-Royce plc)

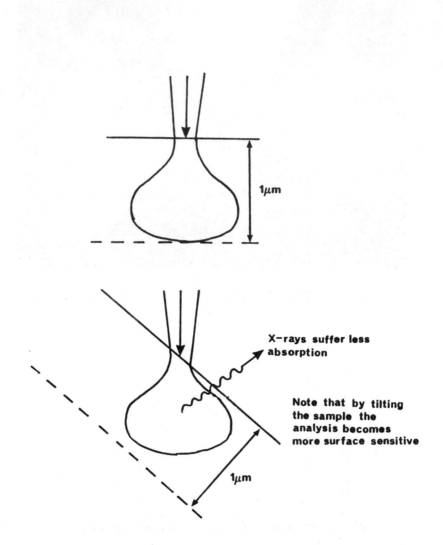

X-rays suffer less
absorption

Note that by tilting
the sample the
analysis becomes
more surface sensitive

Figure 26.Effect of the surface tilt

180

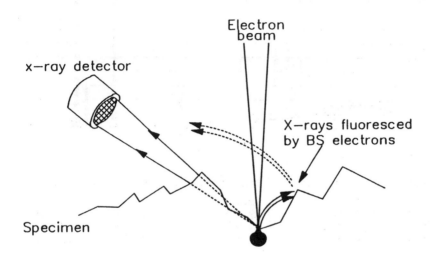

Figure 27.Effect of surface roughness on the collection
of X-rays

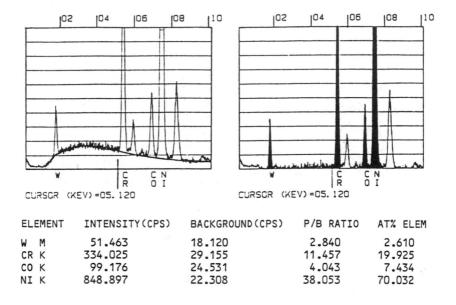

| ELEMENT | INTENSITY(CPS) | BACKGROUND(CPS) | P/B RATIO | AT% ELEM |
|---------|---------------|-----------------|-----------|----------|
| W  M    | 51.463        | 18.120          | 2.840     | 2.610    |
| CR K    | 334.025       | 29.155          | 11.457    | 19.925   |
| CO K    | 99.176        | 24.531          | 4.043     | 7.434    |
| NI K    | 848.897       | 22.308          | 38.053    | 70.032   |

Figure 28.Semi-quantitative analysis of a sample
        containing cobalt, nickel, chromium and
        nickel showing:the original spectrum, the
        spectrum with background subtracted and
        characteristic peaks identified for analysis
        and a table containing results.

CONCENTRATION

WT.%   AT.%   % S.E.

```
W M  11.50   3.90   3.28
CRK  15.90  19.06   0.84
COK   7.06   7.46   1.80
NIK  65.54  69.58   0.50
     ------
     100.00
```

ZAF CORRECTION

ELEM   K     Z      A      F

```
W M 0.063 0.862 0.596 1.000
CRK 0.174 0.999 0.940 1.102
COK 0.070 0.987 0.946 1.010
NIK 0.692 1.025 0.960 1.014
```

MASS ABSORPTION COEFFICIENTS
EMITTER
ABS.

```
        W      CR     CO     NI
W   1444.2   478.4  252.0  206.8
CR  1824.6    87.4  385.5  313.3
CO  2635.4   126.2   64.4   52.3
NI  2948.1   141.2   72.0   58.5
```

Figure 29.The results from a fully quantitative
analysis of the spectrum in fig.28.

# 5 : Wavelength dispersive spectroscopy
# T E ROWEN

T E Rowen is in the Department of Wrought Technology and Process Physics, at Rolls-Royce plc, P.O. Box 31, Derby DE2 8BJ.

## 1  BRIEF HISTORY OF ELECTRON-PROBE MICROANALYSIS

### 1.1  THE EARLY DAYS

In 1821 Michael Faraday discovered the principle of the electric motor and some years later he invented the dynamo for generating electricity. In 1867 Faraday died, but his experiments on electrolysis were extended with an invention called a Crookes Tube or cathode ray tube. The results of these experiments strengthened the notion that electricity existed in discrete units or particles and in 1874 an Irish physicist named Johnstone Stoney coined the name electrons for these particles. They were actually discovered by J J Thompson and for this work he was awarded the Nobel Prize in 1906. The first inkling that the electron played a principal part in determining the chemical and physical properties of the elements originated in the laboratory of Ernest Rutherford in Manchester. While Rutherford was unravelling the structure of the atom in another part of the same laboratory the basis of the microscopic explanation of Mendeleev's periodic table was developed by a brilliant young physicist, H G J Moseley. By studying the X-rays emanating from the atom, Moseley showed that the correct independent variable to use in the periodic table was the atomic number and, in 1911, he stated his famous periodic law.

In modern form, this law states that: "the chemical and physical properties of the elements are periodic functions of the atomic number".

It was in 1912 that Bragg had stated that a crystal could be used as a three-dimensional grating for diffracting X-rays and that:

$$n\lambda = 2d \sin \theta$$

and this famous law, whilst being an integral part of X-ray analysis, became the basis of another analytical technique, X-ray diffraction.

The final part of the jigsaw puzzle that goes to make up electron-probe microanalysis (EPMA) was supplied by de Broglie (2.9) who stated that electrons and photons have a dual personality of energy and wavelike properties and energy and wavelength are related according to the equation:

$$E = hc/\lambda$$

The man who put these facts to work was a PhD student at the University of Paris named R Castaing and he set out in a remarkable thesis (Castaing, 1951) the foundations for the entire technique of EPMA. It was in 1948 that Professor Guinier asked Castaing to build a microscope and make a point-to-point analysis of a metal by bombarding it with electrons and measuring the characteristic X-ray emission. This early microprobe, however, used a Gieger counter which cannot differentiate between wavelengths (Castaing actually used the variation in count rate to analyse binary alloys) but in 1950 the fitting of a crystal spectrometer allowed the possibility of quantitative analysis. Thus in 1951 Castaing laid the foundations for EPMA, although, because of a previous patent (Hillier, 1947) for a microscope which was never built, he is not, in law, the inventor (Mulvey 1983).

## 2 THE PHYSICS OF X-RAY GENERATION

If the reader wishes he or she may omit the following chapter because it has been covered in the previous monograph. For the sake of completeness and easy reference however, the interactions occurring in X-ray generation are described as follows.

### 2.1 ELECTRON INTERACTIONS
The generation of X-rays in a solid can be achieved by

bombarding it, in vacuo, with electrons of sufficiently high energy. In X-ray spectrometry this is done in an electron-probe microanalyser (EPMA) or a scanning electron microscope (SEM) and the accelerating voltage of the electron gun (6.2) is usually 15kV or greater. Here confusion can arise. Some authorities would write 15keV for the accelerating voltage because 1kV at the electron source represents 1keV at the sample. The energy of the X-rays measured at the detector will always be in keV so it is the author's view that the accelerating voltage should be written in kV and that will be the convention adopted here.

## 2.2 BACKSCATTERED ELECTRONS
The interactions which occur between primary or incident electrons and the target atoms can be divided into elastic and inelastic processes. Elastic events occur when the primary electron collides with the relatively massive atomic nuclei when there is little chance of energy interchange and the primary electron "bounces back" with almost the same energy with which it arrived. This electron is then known as a "backscattered electron" (BSE) and provides the basis for atomic number difference imaging using a BSE detector. As shown in Figure 1, BSEs are generated in a small area (compared to the X-ray volume) so giving good spatial resolution and the denser the atomic nuclei the more electrons backscatter from it so the brighter the image. Backscattered electrons can be particularly useful in EPMA when a polished specimen can appear featureless in optical or secondary electron mode (of which more later). Hidden phases can be revealed by a BSE detector, as in Figure 2, thus greatly facilitating the location of analysis points.

## 2.3 INELASTIC SCATTERING
Inelastic scattering results from interaction between the primary electrons and orbital electrons of the atom. (Although these orbitals do not actually exist, for the purpose of this monograph we will use the Bohr model in which electrons live in fixed orbitals). Here the difference between the energy levels of the atom are less than the energy levels of the primary electron and energy transfer readily occurs. In other words an orbital electron is knocked out of its orbital and becomes a secondary electron. The point of importance here is that secondary electrons have low energies, typically a few electron volts (eV) and only those close to the surface can escape and, when collected

186

with a secondary electron detector (SED), are used to form high-resolution images on the cathode ray tube (CRT) of the EPMA or SEM. These secondary electrons are highly sensitive to the topography of the surface (Figure 3) and this, and their high spatial resolution, make them the most frequent choice for micrographic images.

When the atom is in an excited state having lost an inner orbital electron it can be referred to as an excited ion and will give up energy to return to a normal ground state through a relaxation, or de-excitation process. The most likely process is a series of transformations in each of which an electron from an outer shell "drops into" a vacancy in an inner shell. Each drop results in the loss of a specific amount of energy, namely, the difference in energy between the vacant shell and the shell contributing the electron. This energy is given up in the form of electromagnetic radiation; X-rays in the case of high-energy transitions involving inner shells and the energy of the radiation is uniquely related to the element from which it emanated according to Moseley's (1911) famous periodic law.

## 2.4 X-RAY LINES
A discussion here of the nomenclature for X-ray emissions might be helpful. The results of X-ray emissions are usually viewed as lines (or peaks) on a spectrum and the lines are named according to the shell in which the initial vacancy occurs and the shell from which the electron drops to fill that vacancy. See Figure 4. Thus if an initial vacancy occurs in the K-shell and the shell from which the electron drops to fill that shell is an adjacent shell (the L-shell) a $K\alpha$ X-ray is emitted. If the electron drops from the M-shell (two shells away) the emitted X-ray is a $K\beta$ X-ray. Those empty shells can now be filled with electrons falling from M or N shells with their own X-ray emissions, according to the Pauli exclusion principle. Figure 5 is a schematic of the electron shell arrangement for the whole periodic table and from this it can be seen that with atoms containing a lot of electrons, many interchanges can occur and many X-ray lines can be generated.

## 2.5 AUGER ELECTRONS
There is an alternative de-excitation route for the atom to take and that is for the X-ray photon emitted from, say, the K-shell to be absorbed completely by an

187

L-shell electron and that electron be ejected from the atom. This is an Auger electron and itself possesses an energy exactly equal to the difference between the energy of the original characteristic X-ray and the binding energy of the ejected electron. In contrast to characteristic X-rays, Auger electrons are of very low energy and travel only a short distance in the sample, so can only escape from the first few atom layers. Auger analysis is a discipline in its own right and is a true surface analysis technique.

## 2.6 FLUORESCENT YIELD
The probability of an X-ray photon rather than an Auger electron being produced is called the fluorescent yield. It increases with atomic number and is larger for K-line emissions than L-line emissions as in fig.6.

The low fluorescent yield for the low atomic number elements is one of the factors limiting their detection. For example, the fluorescent yield of the K-shell decreases from ~0.8 for elements Z=40, to ~0.1 for element Z=16; in the ultra-light element region fluorescent yields are very low, ~0.001 for carbon X-rays.

## 2.7 IONISATION POTENTIAL AND OVERVOLTAGE
For an atom to be ionised the energy of the incident electron must be greater than the critical excitation potential of that atom. It is considered that each shell of electrons has an absorption edge, or critical excitation potential which is a value up to which absorption of the incident electron will take place but above which ejection of an electron will occur. Each orbital energy has a discrete value and K-electrons require greater energy than L-electrons as shown in figure 7.

Since the critical excitation energy ($E_{crit}$) is the energy required to remove the orbital electron from its shell, it is approximately equal the total sum of the energy orbitals outside that shell. Thus the $E_{crit}$ of tungsten (69.5keV) is equal to the sum of the K+L+M energies (59.3+8.3+1.7) for that element. It is not sufficient however, to simply exceed the $E_{crit}$ value for the element(s) of interest because this would produce very few X-rays. It is usual to exceed $E_{crit}$ by 2-3 times (2.7 being the rule-of-thumb) and, because 1keV at the sample equals 1kV at the anode (2.2), the accelerating voltage in kV is required to be 2-3 times the elemental voltage in keV. Now, if the $E_{crit}$ of tungsten is ~70keV, then 2.7 times that is ~189kV.

188

This poses a problem as most SEMs and EPMAs will not exceed 50kV.

The answer to the problem is to ignore the K-line and analyse the L- and M-lines using a suitable voltage (25kV). This is not a problem in WDS because of its superior resolution, but because L-lines from heavy elements lie very close to K-lines of other elements present in many high-temperature alloys, is a major problem in EDS analysis.

## 2.8 CONTINUUM RADIATION
This is white radiation and is a mixture of all X-ray energies from the primary electron energy downwards and is analogous to white light. Continuum radiation is also known as Bremsstralung (from the German meaning "braking radiation") and is caused by near misses by the primary electron as it passes through the atomic structure which cause it to slow down giving up its energy gradually and continuously and is therefore not characteristic.

## 2.9 SECONDARY FLUORESCENCE
Secondary fluorescence can occur when a characteristic X-ray is absorbed by an adjacent atom which is excited and subsequently emits its own characteristic X-ray photon. The effect is particularly prevalent in the transition elements where the rule of thumb is that elements two atomic numbers apart can cause secondary fluorescence (lower than $Z=11$ the rule changes to one atomic number apart).

The reason is a function of the energy gap between adjacent periodic atoms and in the transition range an iron $K\alpha$ X-ray photon has an energy of 6.4kV. This is a convenient energy to ionise chromium ($E_{crit}$ 5.98keV) and produce characteristic chromium $K\alpha$ x-rays at 5.41 keV, but manganese, which requires an energy input of 6.53 keV cannot be ionised by characteristic iron X-rays. The net result of all this is that the detector collects more chromium X-rays than were originally generated and fewer iron X-rays than were originally generated.

In quantitative analysis complex correction procedures take account of this effect (5.7).

## 2.10 RELATIONSHIP OF ENERGY TO WAVELENGTH
In the last few paragraphs we have discussed X-ray energies when the title of the monograph is wavelength dispersive spectrometry. This apparent anomaly can be explained by the work of Louis de Broglie who

promulgated the idea that electrons had particle and wavelike characteristics by re-writing Einstein's famous equation:

$$E = mc^2$$

as

$$E = hc/\lambda$$

where h is Plank's constant, c is the speed of light, E is energy and $\lambda$ (lambda) is wavelength.

So, if an electron or photon has energy it must have wavelength and the relationship is:-

$$E \text{ (keV)} = 1.24/\lambda \quad \text{(nm)}$$

Finally, as will be obvious, hydrogen and helium have too few electrons to produce X-radiation and lithium and beryllium produce X-radiation too soft (or too long in wavelength terminology ; $BeK\alpha$ = 6.76nm) to pass through the windows (3.8) in an EPMA so the lightest element which can be detected is boron (Z=5).

## 3 SPECTROMETERS

### 3.1 THE CRYSTAL SPECTROMETER

A wavelength dispersive spectrometer is a mechanically driven device which consists of a bent and ground crystal monochromator, a gas filled detector and a single channel analyser. Modern EPMAs feature fully focusing spectrometers in which specimen crystal and detector lie on the same focusing circle, known as the Rowland circle. The X-rays emitted by the sample are analysed by the monochromator and, if this is set at the Bragg angle for a given wavelength, the X-rays are reflected through the detector window and counted. The total arrangement is show in Figure 8.

The monochromator uses the ability of crystals to selectively reflect X-rays from a family of reflecting planes. The reflection obeys classical laws of optics; the angle of reflection is equal to the angle of incidence so we are able to detect reflected X-rays only if the successive reflections from different planes remain in phase. The path difference travelled must therefore be an integer multiple of the wavelength. This requirement is stated in Bragg's Law which relates the wavelength, lambda ( $\lambda$ ) to the interplanar spacing d, and the angle of incidence theta ($\theta$) and can be written:

$$n\lambda = 2d \sin\theta$$

when n is the order of reflection

Figure 9 shows the schematic of crystal diffraction and we assume that the scattering centres are points. In reality, the atom is composed of an electron cloud of variable density. There is therefore, a structure factor for each atom which expresses the intensity of radiation scattered in the 0 direction. This means that diffraction of higher order reflections will differ depending on the nature of the crystals. For some, like quartz, the diffracted beam intensity decreases rapidly as n increases; for others, such as mica, it decreases very slowly. Some crystals such as PET (pentaerythricol) preferentially attenuate even numbered orders. Despite the fact that spectral resolution is better for higher orders, microanalysts prefer first order spectra and thus prefer crystals like quartz or PET to mica.


## 3.2 ULTRA-LIGHT ELEMENT CRYSTALS

Until recently the crystals used for ultra-light element ($z=<10$) have been manufactured from lead stearate (known as ODPb, lead octodecanoate) which are layers of heavy metal cations separated by chains of organic acid. The cations act as diffracting layers and the size of the organic chain controls the d-spacing.

New multilayer devices which are now commercially available consist of alternate layers of high and low atomic number element deposited upon a smooth substrate by sputtering or evaporation. Their diffraction efficiency may be optimised by appropriate selection of the deposited elements and any d-spacing can be chosen (Love & Scott: Progress in the EPMA of light elements). For carbon analysis, for instance, a nickel-carbon multilayer (sometimes known as ML2) with a 2d spacing of 9.35nm is available. Its resolution is actually poorer than ODPb but it scores heavily on peak-to-background ratios and provides much higher intensities than stearates, being ~3 times more sensitive to carbon X-rays than ODPb. This means that minimum detection limits are greatly improved and the better peak-to-background ratios provided suggest quantitative analysis of light elements may now be practicable. Table 1 shows some common crystals and their characteristics.

## 3.3 RESOLUTION

The essential feature of the crystal spectrometer is its resolution, that is, its ability to resolve two lines very close to each other. The resolution is defined as the width of the line measured at full width half maximum (fwhm) and is governed by a number of factors including crystal perfection, Bragg angle, focusing conditions etc. In WDS the actual resolution is not often measured though a typical peak width may be of the order of 10-3nm which is equivalent to an energy resolution of between 5 and 50eV. This resolution is superior to that of an EDS detector which exceeds 100eV.

Crystals can degrade with constant X-ray bombardment but the effect is small, gradual and difficult to monitor. The greatest degrader of resolution in both WDS and EDS systems is thermal agitation and the greatest offender is room temperature. Microscope rooms should be kept cool and at a constant temperature if the best results are to be obtained from the spectrometer to say nothing of the microscopist.

## 3.4 REPRODUCIBILITY OF ANALYSIS

We can see from Table 1 that different crystals are needed to cover different X-ray wavelengths but if we study Figure 10 we can see that some crystals overlap and the microscopist is therefore able to choose which ones to use. In real life this will be a compromise based on the number of spectrometers on the microscope, the number of crystals per spectrometer and the number of elements in the sample. This brings us to the point of reproducibility of the spectrometer. With a modern fully focusing crystal arrangement the crystal and detector move mechanically round the Rowland circle as the Bragg angle changes when the element (wavelength) changes. Moving the crystal is achieved by a motor mounted on a shaft to which the crystal is affixed and moving the detector is achieved by a clockwork mechanism the whole of which fits in a large casing which is, of course, evacuated. So if we have nine elements to analyse equally divided between three spectrometers (one crystal each) the mechanism on each spectrometer must move 9 times because for each element (peak) it measures it must also measure the background either side of each peak. Clearly then for 100 analysis points for 9 elements each mechanism must move 900 times, perhaps tens of thousands in a week. This amounts to millions of movements in the life of an EPMA, so only the highest engineering standards and the best materials available will ensure that an instrument can maintain its reproducibility throughout its lifetime.

## 3.5 SPECTROMETER CONFIGURATION

The microscopist may have one further choice to make, because there are two spectrometer configurations, vertical and inclined, which relate to the orientation of the focusing plane with respect to the plane of the sample. The major drawback of the inclined spectrometer is that it is cumbersome, which usually limits it to a maximum of two. This type of spectrometer has the advantage of being relatively insensitive to variations in Z height of the sample (fig.11) which can be useful in the analysis of inherently undulating specimens (4.1). In contrast vertical mounting of the spectrometer takes little room so it is possible to position several (as many as six) around the electron column giving a choice of wavelength ranges and, of course, speeding up the analysis. The vertical spectrometer is, unfortunately, very sensitive to specimen Z height variations and the effective variation in Bragg angle associated with it.

## 3.6 CRYSTAL CONFIGURATION

A typical EPMA configuration might be two vertical and one inclined spectrometer which gives the greatest opportunity of choice or confusion. A final word of warning, we have seen (2.4) that the fluorescent yield increases with increasing atomic number, and this is true, but the X-ray yield count at the detector varies enormously depending on the crystal chosen. Chromium, for instance, can be analysed using LIF and PET (see Figure 10). Using a chromium pure element standard at 15kV accelerating voltage and 25nA absorbed current the X-ray yield in counts per second (cps) would be ~7,000 using LIF but with PET it would be ~3,000 cps. So, if the concentration of chromium in the sample was low the counting statistics could be improved by the choice of crystal.

Conversely for pure titanium using the same beam conditions the LIF crystal gives ~4,500 cps but using PET gives ~35,000 cps. So on a sample containing a lot of titanium, where dead time might be a problem, the choice of crystal could overcome the problem. Of course the resolution might be worse, adding yet another variable to the fascinating science of microprobe analysis (the reasons for the differing count rates quoted above include the crystal reflectivity and Bragg angle).


## 3.7 DETECTORS

The detector is a counter which consists of a cylindrical chamber (cathode) filled with gas, a coaxial metal wire (anode) held at a positive voltage (bias) with respect to the chamber and a window transparent to X-rays and capable of holding a vacuum. The counter may be either sealed or may permit a steady flow of gas, as in Figure 12. In the latter case, the life time of the counter is practically unlimited so long as the wire anode is not contaminated.

When an X-ray photon enters the counting chamber all its energy is absorbed to ionise an atom in the gas by virtue of the photoelectric effect. An electron in the L level is ejected from its orbit (photoelectron) with kinetic energy E1 equal to $E_0 - E_L$ when $E_L$ is the ionisation of the L level. The gas atom is now an excited ion (actually an ion-pair, a photoelectron and a positive ion) and returns to ground state by the process of radioactive transitions with emission of a (fluorescent) photon or an Auger electron. The photon and the most energetic Auger electrons lose their energies progressively by successive inelastic collisions producing other ionisations and the formation of other electron-ion pairs and the total energy lost in ionising the gas equals the initial energy of the incident photon. On average, one ion-pair is created for every 25-30 eV (for argon gas) of incident X-ray energy hence the number of ion-pairs in proportional to the energy of the incident X-ray photon. Under the influence of the radial electrical field between the centre wire and the wall of the chamber the electrons migrate towards the cathode. The number of electrons collected at the anode depends on the applied voltage (+1,000 to 2,000V) and the effect is to accelerate the free electrons created in the initial X-ray absorption such that they cause further ionisations and more electrons (the Townshend "avalanche"). This increased output pulse from the anode wire is still proportional to the incident photon energy and such internal amplification is known as the "gas gain".

When positive ions reach the cathode their neutralisation gives rise to ultraviolet radiation which could produce electrons from the counter wall and disrupt the proportionality. To avoid this a polyatomic may be chosen, such as 10% $CH_4$ bal Ar (10% methane in argon, known commercially as P10) and this addition absorbs the ultra-violet radiation. As already noted, gas counters can be either sealed or flowing and argon methane are the latter variety. Permanently sealed counters usually contain xenon which has a high absorption and is generally preferred for detecting higher energy X-rays.

## 3.8 COUNTER WINDOWS
The window which separates the gas from the vacuum in the spectrometer must be thick enough to prevent leakage but thin enough to permit the passage of low-energy X-rays and it must be made of a low atomic number material to avoid absorption. When dealing with ultra-light element X-rays, it is necessary to lessen the degree of absorption of the gas or most of the primary ionisations occur close to the counter window where the field is too weak to prevent ion-pair re-combination. This is usually achieved by lessening the gas pressure which in turn means a thinner window can be used. Beryllium of thickness 125 μm is commonly used for sealed counters because it can withstand atmospheric pressure without rupture. It will not permit the passage of low atomic number X-rays however and is difficult to make thinner, so plastic windows (Mylar, $C_{10}H_8O_4$, of 6μm or 3μm thickness) are preferred. But for longer wavelengths ( >0.5nm) windows must be made even thinner, from polypropylene (CH2) or collodion ( $C_{12}H_{11}O_{22}N_6$ ) and these thin layers usually require supporting on a grid.

Whatever the windows, they always leak very slightly and eventually rupture. A sealed counter must then be replaced totally whereas with a flowing counter just the window can be replaced. Sealed counters are superior for analysis of high-energy X-rays but cost is generally against them. Flowing counters however, require gas bottles which need replacing before they are empty (to prevent contamination and decreasing gas flow) and these should be in the same temperature-controlled room as the microscope. Argon methane mixtures are, however, considered flammable and so could pose safety problems.

Finally, for ultra-light element X-ray analysis it should be noted that the same rules of absorption apply as in the sample (2.9) and when analysing X-rays from nitrogen it would be best to use a plastic film with a high nitrogen content rather than polypropylene which has a high carbon content which will, of course, absorb nitrogen X-rays. It will be clear by now that a spectrometer which has been designed for detecting lower energy X-rays will not produce optimum performance for higher energy radiations which brings us back to choice. This time though, instead of the choice of operating conditions the microscopist must, when ordering the instrument, specify exactly the type and number of spectrometers required, their crystal configuration, gas types and window thicknesses and material. Many a potentially first-rate instrument has been compromised by an error or an omission in the specification.

### 3.9 ARTEFACTS
Before considering the counting of the electrons at the anode it is worth pointing out that even with a perfect spectrometer arrangement there are certain effects which are unavoidable and these we will note:

**Main Peak:** Even if the counter is operating under ideal conditions the pulses produced by a flux of X-ray photons of a given energy do not all have the same height and the fluctuations arise from the statistical nature of the ionisation phenomenon.

**Resolution:** The resolution of the counter is defined as the width of the peak at full width half maximum (fwhm) and the actual resolution obtained is not as good as the theoretical resolution because of counter geometry, thermal agitation and the anode wire. This should be held taut so it is usually mounted under spring tension; its surface must be perfectly clean. Uniformity of the wire diameter is important for good resolution, so a metal like tungsten with a high yield strength, so it can readily draw but not easily contaminated, is chosen.

**Escape Peak:** The energy of the photon, Eo, is totally recoverable by the process described previously if, after absorption of the incident photon, decay occurs by an Auger transition. If the decay occurs by a radiative transition, an X-ray photon of energy E$\alpha$, characteristic of the gas, is emitted. This radiation has a low probability of being absorbed and may escape from the counter. In this case, the energy $E_o$-E$\alpha$ only is recovered. there are thus two noticeable peaks, on centred on energy $E_o$ (the main peak) and the other on energy $E_o$-E$\alpha$ known as the escape peak. These will of course, vary with the element being detected and the gas being used.

**Dead Time and Linearity:** Quantitative analysis is possible only if the output pulses vary linearly with the X-ray intensity the range of linearity is limited. At low X-ray intensities the response is disturbed by the noise in the counter itself due to stray radiation e.g. cosmic radiation, ambient radioactivity.

At high counting rates, the dead time becomes limiting, as it represents the span-time of the positive space charge in the active zone surrounding the anode. While counting, the arrival of a second pulse will be lost in this, the detector dead time. It is, however, of much less consequence than the much greater dead time of the electronics.

### 3.10 SINGLE CHANNEL ELECTRONICS
The electronic circuits associated with a proportional counter are there to amplify the pulse from the counter and shape it into useable form before counting it and Figure 13 shows typical counting circuitry.

The first stage of the counting electronics is a preamplifier which is positioned near to the detector to minimise the effect of the capacitance of the coaxial cables. The first stage of the preamplifier consists of a field-effect transistor (FET) whose impedance is very high and whose noise level is very low. The current of electrons produced in it charges a capacitor which then delivers a pulse of a few millivolts. (A typical preamplifier produces ~0.1µv per ion pair).

The amplifier transforms the low level signal to a high level signal (a few volts) suitable for processing by the pulse height analyser. It not only amplifies the signal but also transforms it to a Gaussian shape from the exponential decay form it had coming in from the preamplifier; in turn, the signal-to-noise ratio is increased and the degree of pulse overlap occurring at high count rates is reduced. The time constant of the whole circuit is usually 1µs which produces output pulses a few microseconds wide, sufficiently short to allow count rates of $10^5$ per second to be recorded without serious dead time effects being created.

### 3.11 PULSE HEIGHT ANALYSIS (PHA)
The single channel analyser eliminates unwanted pulses and transforms the signal into a logical form compatible with the scalar and integrator. Generally, it consists of two trigger circuits in parallel coupled

to an anticoincidence circuit. The trigger delivers a logic pulse when it receives a signal equal to or greater than a preselected value (threshold value); the anticoincidence circuit continuously compares the signals coming in from the two trigger circuits and delivers a logic pulse only if it receives a single signal. This rectangular logic pulse is recorded by a ratemeter or scaler. The width of the logic pulse contributes to the dead time of the system.

The PHA also contains an upper level discriminator so that a "window" of adjustable width may be used. Only those input pulses whose height exceeds the lower level but not the upper level will generate an output pulse. The window setting is useful for suppressing any higher order reflections from the analysing crystal.

X-ray micrographs showing the distribution in the specimen of an element selected by the spectrometer may be formed using the output directly from the PHA. A scaler counts the arrival of logic pulses in a given time internal (usually 10s). The time interval is measured by a separate scalar which counts pulses generated by an electronic clock. This raw data is then fed to a computer where it will be stored for quantitative analysis using correction procedures.

### 3.12 QUALITATIVE ANALYSIS
The foregoing has been largely devoted to acquiring data, that is, we know what elements to look for and arrange the spectrometers to go to a prearranged setting and record the counts. With a totally unknown sample we would first have to establish which elements were present by scanning the spectrometer through the full range of sin $\theta$ for each crystal (remembering that we have to use at least 4 crystals to cover all the elements attainable). As we have seen, some crystals give a higher fluorescent yield (3.5) than others for the same element and also low concentrations are difficult to detect so we would start by scanning the spectrometers as slowly as possible to give maximum chance of detection. Without being interminably slow, this means about 10 mins (although it is unlikely that we would know absolutely nothing about the sample). So, as the spectrometer scans, the Bragg angle for the crystal alters and whenever an element is present counts are recorded as described previously. A computer records these events as they occur and can match the wavelengths recorded against known data and give a read out of elements present. It will however,

give only its best guess (which usually means some alternatives) and it will be up to the microscopist to sort out which elements are actually there by using look up tables and, if recorded counts are low (double figures) run the spectrometer again, more slowly, over the region of interest.

One of the most often asked questions in EPMA is when is a peak a peak? The answer is when the background count is exceeded by a certain amount. The formula is:

$$3\sqrt{P/B}$$

which means that the smallest detectable peak may be defined as three standard deviations of the background count.

The point is, if the background counts are not exceeded it means a peak may be present but we cannot detect it because it falls below our minimum detection limit (MDL). Whilst the theoretical MDL may be 0.001 wt% (it varies with the element and accelerating voltage) in real life the MDL is more likely to be 0.01wt%.


## 4 SAMPLE PREPARATION

### 4.1 REQUIREMENTS FOR EPMA
Sample preparation for EPMA would appear to be simple. All that is required for quantitative analysis is a flat polished specimen so that the Z height, relative to the crystal, is constant across the specimen which ensures that X-ray absorption is constant from point to point. An example of a high-temperature material is a ceramic, such as zirconia (zirconium oxide) which is used for heat resisting thermal barrier coatings (TBC). The first problem is sectioning the sample prior to mounting in conductive bakelite. Ceramics are hard, brittle and friable and difficult to section without damage so they are often impregnated with some sort of resin before sectioning. This impregnation also assists with the polishing because ceramics (and some metallic alloys) suffer from pieces breaking out of the sample as it is abraded; known as "pluck out". All of this assumes that the grinding media is harder than the ceramic and that the sample does not grind away the polishing equipment. The art of sample preparation is known as metallography and is worthy of a book in its

own right but some of the more common polishing media
are shown below:-

| Materials | Media |
|-----------|-------|
| Aluminium | MgO + H2O |
| Titanium | Al203 + H2O |
| Iron | SiC - Diamond |

The object of the exercise is to use successively finer
polishing media to produce a flat, level, optically
scratch free surface.

Metallic alloys can suffer from pluck out because
they often contain a second phase which gives extra
strength. Such phases are usually intermetallic
compounds, e.g. FeC in steels, $Ni_3Al_2$ in nickel-based
alloys and WC in tungsten containing tool steels.
These materials have a relatively soft matrix with a
relatively hard and brittle second phase which requires
polishing away to the same level and to the same degree
of reflectivity; that is, to a 1μm finish. What this
means is that because the wavelength of (white) light
is $\sim$1μm (the visible spectrum is 7.5 x 10-$^7$ m to 4 x
10-$^7$m) scratches smaller than 1μm cannot be seen
optically. An unfortunate side effect of this
immaculate preparation is that on most EPMAs the Z
height focusing (Fig 11) is achieved using a built in
(fixed magnification) optical microscope which can be
difficult to focus because the specimen has no features
to focus on.

Such specimens are rarely flat, however, and hard
phases usually stand proud of the matrix but to a
tolerable degree, say 0.5μm and some minor surface
imperfection is usually visible to aid focusing.

### 4.3 MAGNETIC SAMPLES
Two other problems associated with metallic and
non-metallic samples are worth considering and the
metallic one is particular to ferritic steels and is,
of course, ferromagnetism. This affects SEM imaging
as well as analysis and in all instruments it causes a
deflection of the electron beam. The exact electron
interactions are beyond the scope of this monograph but
what happens is that with time the electron beam, and
by definition the analysis point, move. The
implications of this when analysing a small phase or
inclusion are obvious; we are not analysing what we are
aiming at. The specimens may be demagnetised to some
extent by de-Gaussing but even then some alloys can be
troublesome.

200

## 4.4 NON-CONDUCTING SAMPLES

The remaining problem with a well polished ceramic sample is that it is non-conducting, it charges up under the electron beam. In EPMA the X-ray source is typically 2 μm in diameter and extends a similar distance below the surface (see Figure 1). Most of the energy of the primary electrons is dissipated in this small volume so the energy density may be quite high; at a primary electron energy of 30kV and a beam current equal to $0.1$nA the energy density at the surface is $3 \times 10^{-9}$ Wm$^{-2}$. This can lead to severe problems because the build up of charge at the surface as the electrical potential at the point of impact rises produces a deflection of the beam and causes specimen current instability with the associated fluctuation in the X-ray count.

## 4.5 CONDUCTING COATINGS

After the great lengths taken to achieve a stable beam current ( 6.2 ) it follows that an unstable specimen cannot be tolerated and the partial remedy is to apply a conducting coating to it. The application of a conducting coating to a specimen with poor electrical and thermal conductivity can alter the X-ray emission from it by changing its backscatter and absorption properties but in practise the effects are small if the coatings are thin, ~10nm. Typical coating materials are carbon, aluminium, copper, gold and palladium and it should be noted that carbon is a poorer conductor than the heavy metals hence a thicker coating should be used. This has little effect on the primary electrons but can be significant in light element analysis due to X-ray absorption. Love and Scott (1983) have noted that such absorption effects will be marked when measuring soft X-rays; for example the mass absorption coefficient of oxygen Kα X-rays in carbon is 12000 cm2g-1 and the fraction lost in a 20nm thick layer is given by

$$\frac{\Delta I}{I} = 0.072 \text{ i.e. the loss is } 7.2\%$$

By contrast the absorption of copper Kα radiation for the conditions cited above is only 0.2%.

## 4.6 COMPATIBILITY

In quantitative EPMA the X-ray count from an unknown specimen will be compared with the X-ray count from a known specimen (standard) so it follows that both will

201

be in same condition so the standard must also be coated with the same element to the same thickness.

This is not easy and is best achieved by coating specimen and standard simultaneously, although reasonable reproducibility may be obtained by introducing a film thickness monitor into the coating equipment.

Specimen coating is usually carried out by either thermal evaporation or by sputtering. Evaporation is favourite as it can cope with all the possible coating elements whereas sputtering produces extremely low yields for carbon and aluminium. Finally, the actual element chosen will be determined by the need to avoid X-rays from the coating interfering with X-ray lines from the sample. Because of the high spectral resolution of the wavelength dispersive spectrometer it is only necessary to ensure that the sample does not contain the coating element. In energy dispersive analysis however one needs to ensure that coating X-ray lines are well separated from the spectral X-rays, or better still, use carbon coating since most ED systems cannot detect elements with $Z<11$.

## 5 QUANTITATIVE ANALYSIS

### 5.1 BACKGROUND POINTS
Quantitative analysis is the area where it is most easy to get into difficulty, especially when discussing the results. Most users can set up the electron column, the sample, spectrometer, accelerating voltage and beam current. Normally, the computer programme for quantitative analysis requests the input of these conditions, the elements to be quantified and the standards to be used. It will also request the input of the background points because the number of counts in any peak are considered to be those above the level of background or continuum. Whilst this may sound straightforward, great care must be taken to ensure that when peaks are very close together the position chosen for measurement reflects the true slope of the background and not the slope of an adjacent peak. To achieve this one can usually plot out the raw data as a spectrum and, by measuring the X-axis (which is the range of the spectrometer) calculate the sin 0 position of the background as in Figure 14.

Some manufacturers provide a software programme whereby when the elements to be analysed are input the computer gives recommended settings for background points but care is still required (because spectra can feature artefacts such as satellite peaks) especially in the ultra-light element region.

## 5.2 STANDARDS
Quantitative analysis is based on Castaing's first approximation which measures the ratio, $k_A$, of the characteristic X-ray emission from the specimen (Ispec) to that from a (pure element) standard (Istd) under identical analysis conditions and:

$$K_A = \frac{I\ Spec}{I\ Std}$$

which gives:

$$K_A = C_A$$

where $C_A$ is the mass concentration of the analysed element (A). In this model it is considered that for each sample, all the incident electrons traverse a trajectory of equal length, maintaining the same efficiency for the whole length. We have seen that this is not true (2.1) so a correction factor must be added. Classically this is a ZAF correction procedure when Z = atomic number (backscattering), A = absorption and F = fluorescence and each can be considered separately for clarity though they are considered integrally in some correction programmes.

It follows that standards are prepared to the same condition as the sample and they will be as similar to the sample as possible. Often standards will be pure elements of 99.99 or 99.999% and they will be stable under the electron beam. In the analysis of sulphur, for instance, we would not put a lump of sulphur under the electron beam (not least of all, we do not want to contaminate the vacuum chamber and column), but would choose a compound standard of iron sulphide ($Fe_2S$) and for chlorine, sodium chloride or rock salt. Clearly for oxygen we must use oxides, and these must be as near to the sample composition as possible, to negate the effects of absorption in the standard. There would be little point in using thorium oxide as an oxygen standard for iron oxide because the oxygen absorption is quite different in each.

Such compound standards must be of known composition which means analysing them by another technique, say wet chemical, before they are suitable for standards for EPMA. Microscope manufacturers and consultants can provide ranges of standards of certified composition for most requirements except for carbides, which are not currently available.

## 5.3 ZAF CORRECTIONS: THE ATOMIC NUMBER

The ZAF corrections will be briefly described, together with the model most often used for each one. In the atomic number correction it is recognised that electrons entering the surface of a sample both penetrate it and are scattered by it. Figure 18 shows the effects of accelerating voltage and atomic number on electron penetration and ionisation.

The effects of these processes on X-ray emission can be considered more simply in terms of two factors R and S. The electrons which are backscattered from the specimen surface do not contribute to X-ray production and the fraction of the incident electrons which enters the sample and remains within it is called R. Specimens of low atomic number produce less backscattering and consequently a higher value for R. The electrons which penetrate the specimen may cause ionisations producing X-rays or they may be scattered within it (2-1). Production of X-rays by ionisation depends on the critical ionisation potential ($E_{crit}$) of the specimen and thus on its composition. Elements of low atomic number have lower critical ionisation potentials, are more easily ionised and have greater stopping power per unit mass.

The stopping power of the specimen is call S and is higher for elements of low atomic number. The initial energy ($E_o$) of the electrons affects the values of R and S also. Higher energy electrons may more readily be backscattered and escape from the sample, producing a lower value for R, while the stopping power of the sample may be lessened producing a lower value for S. Thus we have R and S varying inversely with both Eo and Z (although not smoothly). This is fortunate since the correction to be applied depends on the ratio of R + S i.e.

$$C_z = R/S$$

Thus the two factors tend to cancel each other out as both the atomic number and electron accelerating voltage vary. Although the atomic number correction Cz is minimised in this way it may still represent an important part of the correction procedure and is included in the correction programmes. The most widely used atomic number corrections are those of Duncomb and Reed (1968) and Philibert and Tixier (1968).

## 5.4 ABSORPTION

This correction is the least accurate part of every programme largely because measured data does not agree with Monte Carlo calculations (5.6). X-rays are generated at different depths within a specimen and hence must travel through different distances to leave the sample and they are thus absorbed to different degrees. The correction for this absorption, Ca, takes into account the shape of the ionised volume, the angle at which the incident electrons enter the sample and the angle at which the characteristic X-rays emerge from the sample and the composition of the sample. Most programmes use the simplified Philibert (1963) or the Philibert (1965) expression for absorption. Ultra-light element ($Z = <10$) absorption is particularly difficult to calculate and for this reason few programmes attempt it. The common practice is to measure oxygen by stoichiometry or, more dangerously, by difference.

## 5.5 FLUORESCENCE

This is the least significant correction because only if the characteristic X-rays being generated in the sample have an energy range which falls within the absorption range of another element in the sample will the correction be required. In the discourse on secondary fluorescence (2.9) the example quoted was the fluorescence of chromium Kα X-rays by iron Kα X-rays. This is called K-K fluorescence. Chromium could also be fluoresced by Lα radiation of gadolinium, for example and this is known as L-K fluorescence. The fluorescence from the continuum has been shown to be insignificant and is omitted from most programmes. Heinrichs (1966) tables of numerical data give the mass absorption coefficient for absorbing and fluorescing elements and these are often used in correction programmes.

## 5.6 MONTE-CARLO MODELS

Before we go on, a few words about the Monte Carlo model. This is so called because it is a mathematical technique, involving random numbers which can be employed to provide numerical solutions to certain types of problem, analogous to spinning a roulette wheel. Monte Carlo models predict the random walk an electron will take through a sample of any alloy, at any accelerating voltage, any take-off angle etc. Prior to high-speed computers, empirical data was used from samples of known concentrations. Collecting such data was a formidable task potentially running to thousands of specimens and indeed much of the data was extrapolated from a few hundred acceptable analyses produced quite early on. For example Castaing decided to determine the way in which the intensity of characteristic X-rays was distributed with mass depth in a target. Although time-consuming, Castaing and Descamps (1955) succeeded in experimentally determining intensity distributions (effectively the actual tear-drop shape) for copper, gold and aluminium at an accelerating voltage of 29kV. From such data the absorption factors were calculated and the resulting curves were for many years the only readily available means of calculating absorption effects in microanalysers, which were therefore constrained to operate at an accelerating voltage of 29kV (Mulvey 1983).

Even with high-speed computers, though, Monte Carlo calculations are time-consuming, so they are not actually used in EPMA. They are used to calculate the corrections required in a ZAF programme which the microscopist runs to correct the raw data. It follows that there are several different ZAF programmes compiled by different authorities and that the more accurate they are the longer they take to do the calculation.

## 5.7 CORRECTION PROGRAMMES

One of the most rigorous ZAF correction programmes is COR2 written by Henoc et al (1973) Henoc (1974). The less rigorous programmes omit some corrections to achieve more rapid calculations on minicomputers, for instance the programme MAGIC IV (Colby 1971) omits the correction for fluorescence by the continuum. This is not unreasonable as the effect is small.

Having applied the correction to the equation

$$K_A = \frac{I \text{ Spec}}{I \text{ Std}}$$

the answer yields a second approximation of elemental concentration. This improved approximation then serves as a basis for recalculation of the corrections and this process is known as iteration. With each iteration the estimate changes less and less, and the necessary net corrections grow smaller and smaller. The results thus converge to a value that reflects the actual concentrations; or at least, until a convergence test applied after each iteration loop has been satisfied or n iterations have been completed (n = 4).

A knowledge of statistics is useful in understanding microprobe analysis results because the question always asked is "how accurate is the result?" A few traps worth avoiding can be noted. Many programmes normalise the results; this is unavoidable if measuring atomic percent but not if measuring weight percent which is more usual in material science. Because the emission of X-ray photons is a random event each analysis point will give a different total result and will not, probably, add up to 100%. If we have 70% totals, then something is wrong; an element omitted perhaps or incorrect k-ratios or standards, whereas totals from 97-103% are acceptable.

Comparing EPMA results with results from other analysis techniques can be fruitful but beware of justifying them, they can both be correct.

Beware of putting values of accuracy on results. There are several text books which give the methods of calculating accuracies and the summer school book of Love & Scott (Quantitative Electron Probe Microanalysis, published by Ellis Horwood Ltd) is a good example. Each element in the problem has its own relative error (generally increasing with decreasing Z), error of precision (repeatability) and accuracy (compared to a known standard). The only sure fire way to protect oneself is buy or borrow standards of known (assayed) composition which, when analysed, will allow accuracies to be determined.

## 6 THE ADVANTAGES OF THE WAVELENGTH SPECTROMETER

### 6.1 DEAD TIME
The differences between WDS and EDS are shown in table 3 (after Chandler 1977) and can be summarised here. If ultimate detection limits are required WDS wins. If speed is required and if a number of elements are being analysed the EDS system wins. A typical counting time for an EDS system is 100 live seconds. A sensible count rate (X-ray photons arriving at the solid state detector), of 4000 counts per second (cps) will give 20% dead time, that is, 20% of the counts are lost having arrived when the system was still counting the previous pulse and known as the "unreceptive period". Extendable dead time is usually chosen for EDS such that the arrival of a pulse during the unreceptive period affects the dead time by introducing its own period of paralysis then: live time = clock time plus

dead time (which varies with count rate). So the example above would take 120 clock seconds (but during this time would collect the whole spectrum) in contrast to the dead time in WDS analysis.

Here the dead time is much less (3.9, 3.10) and is generally non-extendable with a counting time of 10 secs, but each element (analysed individually) will have its own dead time which will vary according to X-ray intensity. Counting times of 10 secs are usual in EPMA, but where X-ray intensities are low, such as trace element analysis or using a WD spectrometer on an SEM, 100 secs may be necessary.

## 6.2 DEFINITION OF AN EPMA

To clarify the last point it will be necessary to define the difference between an EPMA and an SEM. Before doing that it is worth mentioning that there are three general categories of modern electron column. We have mentioned transmission electron microscopes (2.6) which analyse thin specimens and have high spatial resolution, but they are rarely fitted with WDS so no further mention will be made of them here.

Scanning electron microscopes (SEM) are in wide general use and are designed to give images of high spatial resolution, usually using the secondary electron signal. High spatial resolution by definition means a small spot size, thus a small ionised volume and so a low X-ray yield.

The image displayed on a cathode-ray tube (CRT) is created by scanning the focused electron beam in a raster patter across an area of the sample whilst synchronously scanning an analogous pattern on the CRT. SEMs can vary their accelerating voltage from 100V to 50kV and most instruments use analogue electronics. The latest generation of SEMs, however, are totally digital, software driven and can give automatic control of focusing, astigmatism correction and brightness and contrast.

An electron probe microanalyser (EPMA) is essentially an electron column designed to deliver stable beam currents of high intensity; it will have scanning and imaging facilities and the latest generation are fully digital. It is the column, however, which defines an EPMA and is worthy of further description. It may feature a special column liner to prevent stray electron interference and must have a beam regulation aperture with a stabilising electronic feedback    This is most necessary because the emission of X-ray photons from a sample is a random occurrence

208

and the only control of it which the microanalyst has is of the number of primary electrons arriving at the sample per second per unit area. The beam current stabiliser reads the annular portion of the electron beam on the probe forming aperture. This current reading is then used to control the focal length of the condenser lens through the beam stabilising electronic feedback and the power supply of the lens. When the beam current varies the condenser compensates automatically for the change thus achieving a stability better than 1% over a 12hr period. Figure 17 shows the configuration of the beam stabilisation device fitted to the Cameca Camebax Micro EPMA (courtesy Cameca U.K.).

This measuring aperture should not be confused with the final aperture which is not usually used during analytical EPMA. (It may be inserted for high resolution microscopy and the taking of micrographs).

This means the final lens can run full bore and achieve the high beam current necessary to achieve the required count rate. Running without a final aperture naturally means a relatively large spot size. Manufacturing literature suggests the following values at 30kV.

| Beam Current/amp | Probe diameter/micron |
|---|---|
| $10^{-6}$ | 0.70 |
| -7 | 0.30 |
| -8 | 0.12 |
| -9 | 0.055 |

A typical beam current measured at the specimen (the absorbed current) for a metallic sample would be ~25nA at 15kV.

The above conditions can only be achieved with the electron gun adjusted in a certain manner. The gun is basically a heated tungsten hairpin filament (cathode) in a triode configuration and the primary electrons are pulled from it by a Wernelt cylinder which surrounds it. They are then accelerated by an anode and the characteristics required for EPMA are governed by the filament to Wernelt to anode ratio. (The reader may find details of this process in any standard reference work). Simply put, the nearer the filament is to the Wernelt the brighter, less stable and shorter lived the filament will be. Thus EPMA filaments are set well back from the Wernelt, beams are dull, stable and filament lives can easily achieve 400 hrs in typical conditions. Whilst the conditions in an SEM are the opposite to achieve high spatial resolution they are still often fitted with WDS detectors. Possible configurations are shown below:

| EPMA | SEM |
|------|-----|
| WDS only | EDS only |
| WDS and EDS | EDS and WDS |
| EDS only (unlikely) | WDS only (unlikely) |

Figure 18 shows the EPMA at Rolls-Royce plc Aero Engine Division in Derby. The configuration is of two vertical and one inclined wavelength spectrometers plus a Kevex energy dispersive spectrometer.

It follows then that when doing analysis on an SEM using WDS, it is often a problem getting enough beam current and the beam may not be stable. The problem of beam current is usually overcome by increasing the spot size which achieves the necessary counts, but the inherent instability cannot be overcome. It can, however, be monitored and corrected for at the end of the analysis. Monitoring the absorbed current is one method and some SEMs have provision for inserting a Faraday cup into the electron beam path so the current can be read, say, every 10 analysis points.

## 6.3 MIXED SYSTEMS
Because quantitative light and ultra-light element analysis can only be done by WDS some SEM users have a mixed WDS/EDS system with one wavelength spectrometer controlled by the EDS system. Most EDS manufacturers can provide software for this function and in general the systems work well. The wavelength spectrometer is often inclined and contains several crystals so for maximum flexibility the software should be able to do a crystal "flip" that is, to swap crystals and select the correct sin $\theta$ setting for different elements at each analysis point. There can be problems however, because as we have seen WDS requires a high beam current to achieve the necessary X-ray yield but such a yield can swamp an EDS detector creating excessive dead time and prolonging the time required for the analysis.

## 6.4 THE MAIN ADVANTAGES OF WDS
As we have seen (3.3) the wavelength spectrometer moves the crystals mechanically from one setting to another and this is time-consuming especially when perhaps two settings are at either end of the spectrometer range. Were it not for this (given a reasonable number of spectrometers), the shorter acquisition time would put

210

WDS on par with an EDS system for speed, and, of course, it is already superior in terms of peak to background ratios and MDLs. Until recently the motors which drove the spectrometers (and the microscope stage) from one setting to another were stepping motors. Such motors receive a pulse of electricity for each step of a predetermined distance. Their progress is measured by mechanical counters which are not reliable at high stepping rates which constrains their speed and to ensure accuracy the motors must backlash at every halt. More recently, EPMAs have been available with linear DC motors which, instead of stepping receive a pulse of electricity sufficient to drive them a distance which is optically monitored very accurately (+ or - 0.5μm) so speed is no longer restricted and the motors move as fast as is mechanically possible and stop without the need to backlash.

This is the reason why WDS is now on par with or ahead of EDS in all respects other than cost.

## 6.5 COMPATIBILITY
The solid state detector and the crystal spectrometer are now being seen, hopefully, as complimentary rather than rival techniques. They should not be used in exactly the same way because their respective advantages and disadvantages make them suitable for different purposes. The principal advantages of the solid state detector lie in its rapid simultaneous acquisition, its high detector efficiency and its ability to analyse rough surfaces.

In contrast its poorer spectral resolution and poor signal-to-noise ratio make it no match in performance for the wavelength spectrometer when quantitative analysis and trace level elemental analysis is required. Lastly, the solid state detector does not permit very low-energy X-ray analysis except in its windowless form when it can, under suitable conditions, detect ultra-light elements but quantitative analysis is not yet routinely available.

## 7 CHARACTERISATION OF HIGH-TEMPERATURE ALLOYS

### 7.1 EXAMPLES: SEGREGATION
A typical high-temperature alloy would be any of the family of wrought nickel-based alloys, sometimes known as superalloys, as used for disc material in the turbine section of jet aero engine. These alloys can contain chromium, cobalt, titanium, aluminium and

tungsten. The process route for such an alloy is to cast material into a suitably sized ingot which is then forged to billet shape and machined to exact shape. Before the machining, however, certain heat treatments must be applied to give the particular characteristics required for the chosen application. These heat treatments will be governed by the size, loadings and expected temperature of the component and its history (forging temperature) but they all seek the same end.

This is to produce a microstructure which will give the optimum properties for the optimum life of the component. Such microstructures are based on the principle of precipitation hardening where intermetallic compound known as gamma prime is precipitated from the alloy and forms a regular structure throughout the alloy and restrains the movement of dislocations, thus giving the required material properties. Gamma prime has the composition $Ni_3Al/Ti_2$ and the matrix material (known as gamma) is a mixture of the other elements.

Unfortunately the alloy may not be homogenous and one example of this is segregation. This is caused by variations in the rate of cooling of the material and is often seen as a macro effect revealed with the naked eye, or, at most, binocular inspection, as changes in grain size.

Such segregation can lead to the incomplete or incorrect formation of the precipitate, which can alter the material characteristics locally, often with disastrous results.

It is essential, therefore, to understand the nature of the segregation and predict the effect it may have on material properties and possibly avoid scrapping the part.

Two methods of determining the elemental segregation are as follows:-

**Linescan:** Having determined the composition of the alloy, the spectrometer wavelengths are selected and the first analysis point taken in the good material adjacent to the segregation. The stage of the microscope can then be moved in small steps so the area of segregation passes under the electron beam and at each step the specimen is analysed. If this is continued through to the good material the result can be plotted out as a graph showing distance (stops) on the abscissa and quantitative results on the ordinate. The microstructure in Figure 19 shows such segregation and Figure 20 shows the graph of elemental concentration as a gradual increase in titanium and zirconium. The increase in titanium suggests an increase in gamma prime density which leads to increased tensile properties (and lower material properties) and on this evidence the material could be rejected.

(A sensible investigator would look for additional evidence to back his analysis. In this case, two simple tests would be a hardness check or etching away the gamma which would leave the gamma prime in relief for it to be measured using an SEM.)

**Area Mapping:** There are several ways of mapping a selected area and the one described here is not fully quantitative but more the sort of technique used in production control, where a rapid turn round of results is required. Firstly, a known standard of homogenous material is acquired and the elements most likely to segregate or having the most influence on properties are analysed and the raw data, the count rate from each spectrometer, are recorded. This needs to be done several times.

The unknown sample is then analysed using exactly the same conditions (with special attention to beam current stability) and, if the compared results differ from the standard results by more than the amount they differ from each other segregation may have occurred. A number of points over an area can be measured in this way so building up a map of elemental concentration/variation.

### 7.2 COATINGS

Another example of the use of precipitation-hardened nickel-base alloys in the aero engine is in cast turbine blade material which contains a greater proportion of gamma prime than wrought disc material.

Because of the high temperature ($\sim 1000^\circ$C) seen by a turbine blade it must be protected with an oxidation resistant coating. Conventionally, this is achieved by diffusing extra aluminium into the surface creating an aluminium rich layer 25µm deep which is known as a nickel aluminide coating. It is important to know the elemental concentration of the coating because when it contains fewer than 17wt% of aluminium its structure is gamma prime and is non-protective. At higher aluminium concentrations (up to 40wt%) the structure is NiAl (beta) and confers oxidation resistance. By using EPMA to monitor elemental concentrations (by stepping the sample. as in 7.1) before and after simple oxidation

213

tests at various temperatures, we can predict the life of the turbine blade coating. Figure 21 shows the elemental concentrations of a newly coated turbine blade and figure 22 shows its microstructure. The concentration of elements at point 150 on the graph is the carbide layer composed of elements insoluble in nickel aluminide and denotes the boundary of the coating.

## 8    PHILOSOPHY

### 8.1    EXPERIENCE
There is little use pretending X-ray microanalysis is an easy procedure. Those who have been practising electron microscopists for any length of time will know that many hidden snags and pitfalls lie in wait to trap the unwary.

An investigator with a good fundamental knowledge of electron microscopy will have little difficulty in grasping the basic principles of X-ray microanalysis and becoming familiar with the topics described here. Certainly a person with a background in physics has a great advantage in appreciating the mode of operation of the instrument. It is a technique which requires a good deal of experience and a certain amount of artifice to perfect.

In many instances such as large industrial companies, the investigator may be asked to analyse many different materials, e.g. steels, nickel, titanium and aluminium based alloys, ceramics and silicon nitrides, coatings, bonded structures and tribological samples. All of these have their own particular set of problems and must be regarded accordingly.

### 8.2    USER GROUPS
Detailed analysis of such samples requires that the microanalyst should be familiar with the nature of the specimen and the effect on it of the electron beam; the operating conditions and their effect on the data; the mathematics involved in interpreting the data, and not least of all, the ability to relate the information to the structural information contained in the image. Such people are rare and more frequently several specialists may share the necessary experience.

User groups and microanalytical societies can provide the very useful function of introducing specialists in one area with others of different area. It is very useful when faced with a totally new problem to be able to call on someone who has tackled similar

problems and can point out the difficulties to be
encountered. This can save hours of microscope time
(not to mention nervous tension) and participation in
such groups should be encouraged.

## 8.3 FINAL RESULTS
It is worth going to some trouble to present the
results of an analysis in a professional manner.
Micrographs of specimens which show surface scratches
or are blurred or astigmatic can spoil an otherwise
well-written report. Hand drawn graphs or printed
tables can look clumsy when compared with those
produced on a high-resolution plotter or word
processor. Try to remember that a report may
ultimately be read by someone who has no knowledge of
your discipline and its difficulties; the final report
of a few pages may be all you have to show for days of
work so it is worth spending a little extra time to
ensure that the report reflects the professionalism
with which you have conducted the investigation.

## 9 GENERAL REFERENCES

1.  S Newsam "An Invitation to Chemistry" 1978 New York
    W W Norton & Co Inc.

2.  "Energy Dispersive X-Ray Microanalysis An
    Introduction (ed D Vaughan) 1983 Foster City,
    California, Kevex Corporation.

3.  "Quantitative Electron-Probe Microanalysis (ed V D
    Scott & G Love) 1983 Chichester, Ellis Horwood
    Ltd.

4.  J A Chandler "X-Ray Microanalysis in the Electron
    Microscope (ed Audrey M Glauert) 1977 Amsterdam
    North Holland Publishing Company.

5.  "Microanalysis and Scanning Electron Microscopy"
    (ed F Maurice, L Merry R Tixier) 1980 ORSAY Les
    Editions de Physique.

6.  "Kevex Analyst Number 6" (ed D Vaughan) 1983 Foster
    City, California, Kevex Corporation.

7   S J B Reed "Electron Microprobe Analysis".
    Cambridge University Press

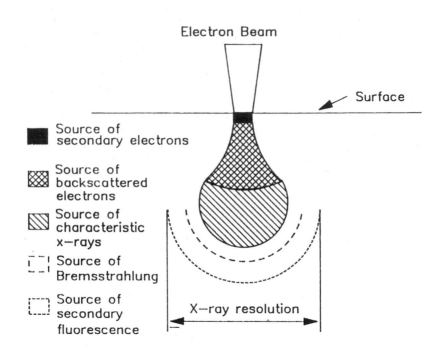

Figure 1. A generalised view of the interaction volumes
and resolution during X-ray generation.

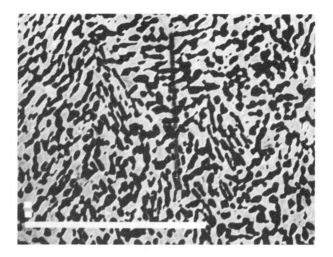

Figure 2. Example of a backscattered electron image.

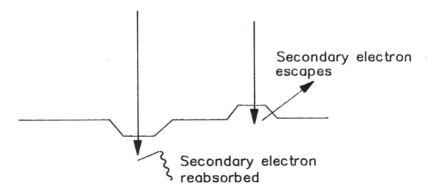

Figure 3. Showing the surface sensitivity of secondary electrons.

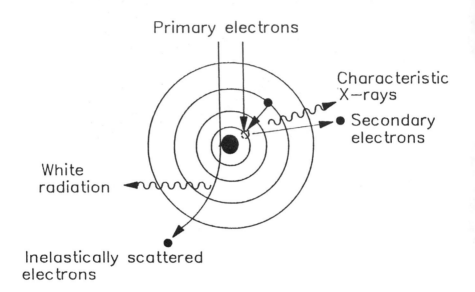

Primary electrons

Characteristic
X—rays

Secondary
electrons

White
radiation

Inelastically scattered
electrons

Figure 4. The classical model of the atom and some of
the signals which are generated during
ionisation.

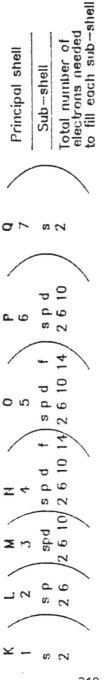

Figure 5. A schematic of the electron shell arrangement for the periodic table and its nomenclature.

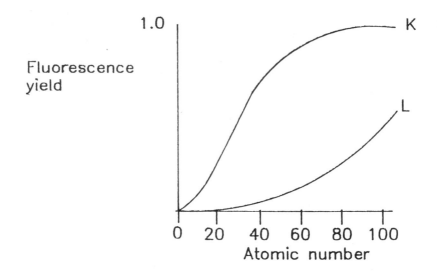

Fluorescence yield

1.0

K

L

0   20   40   60   80   100
Atomic number

Figure 6. Showing the variation of fluorescent yield with atomic number.

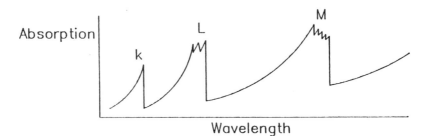

Absorption

k

L

M

Wavelength

Figure 7. Showing K,L and M absorption edges for a given absorbing element.

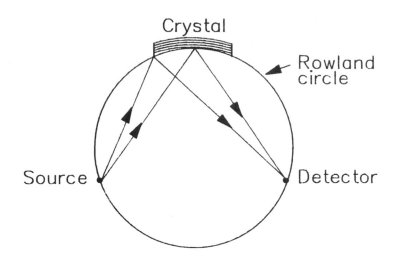

Figure 8. The focusing arrangement of a modern crystal
spectrometer.

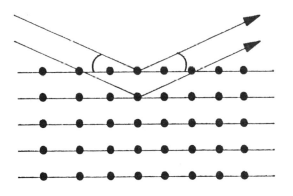

Figure 9. The Bragg reflection.

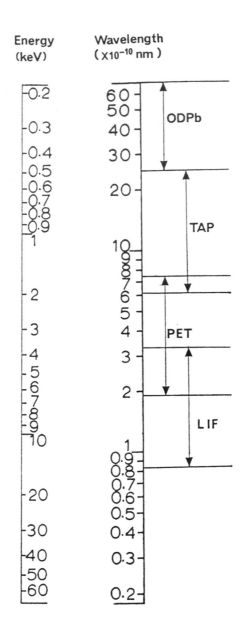

Figure 10. Showing the relationship between wavelength and energy and the manner in which certain crystal ranges can overlap.

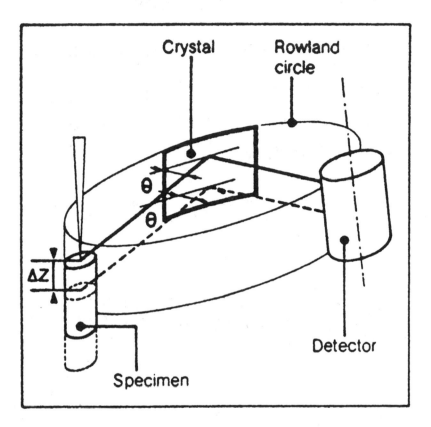

Figure 11. Showing that a change in Z-height in the position of the X-ray source has little effect on the Bragg angle and hence very little effect on spectrometer sensitivity. (Courtesy Cameca U.K.)

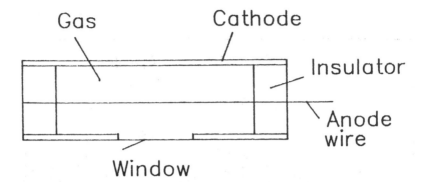

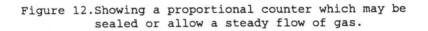

Figure 12.Showing a proportional counter which may be
sealed or allow a steady flow of gas.

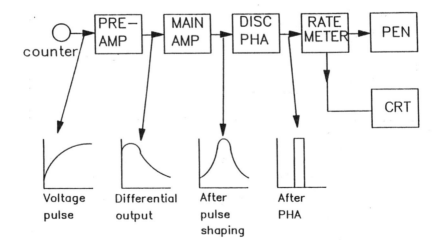

Figure 13.Typical counting circuitry used with a gas
proportional counter.

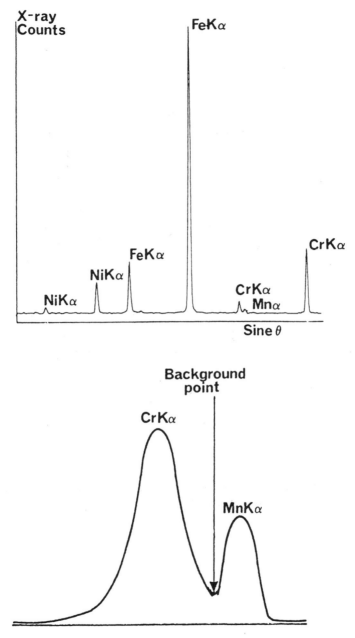

Figure 14. Showing background point selection.

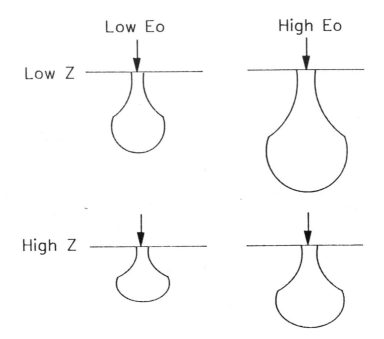

Figure 15.Showing the effect of accelerating voltage on
          ionisation volume.

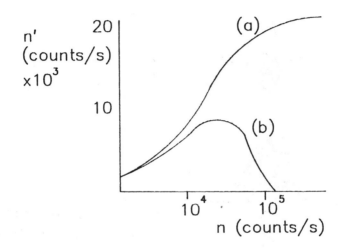

Figure 16.The recorded count rate ,n', versus input
count rate ,n, for non-extendable (a) and
extendable (b) dead time.

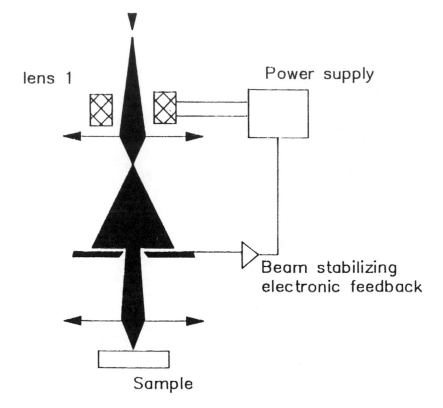

lens 1

Power supply

Beam stabilizing
electronic feedback

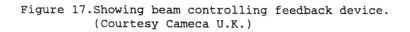

Sample

Figure 17.Showing beam controlling feedback device.
(Courtesy Cameca U.K.)

Figure 18.The EPMA at Rolls-Royce plc.
(Courtesy Rolls-Royce plc)

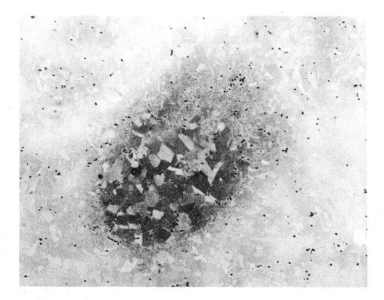

Figure 19.Segregation in a nickel-based alloy.

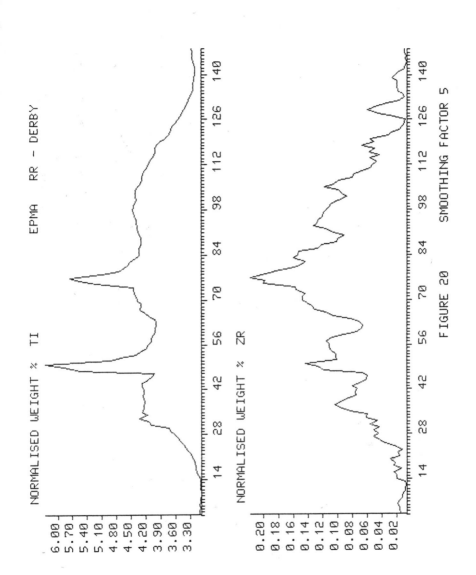

Figure 20.Elemental analysis of segregation in a nickel-
based alloy.

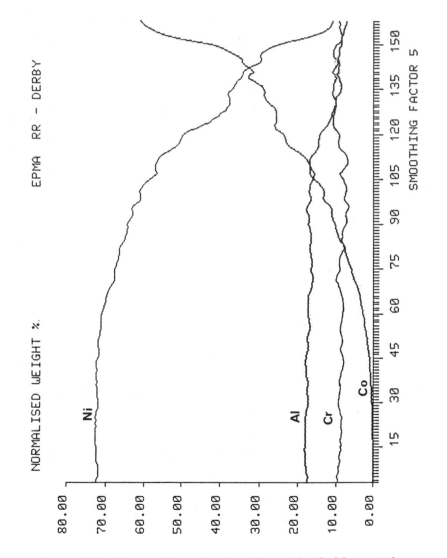

Figure 21.Elemental analysis of an aluminide coating.

Figure 22 An oxidation resistant aluminide coating
(a) aluminide, (b) Ni plate

## TABLE 1

| CRYSTAL | Reflecting Planes | Range | Atomic Number |
|---|---|---|---|
| Lithium Fluoride (LIF) | 200 | 0.1-0.4 | 19-35 |
| Ortho-phthalate rubidium hydrogen (RbAP) | 001 | 0.2-1.8 | 11-14 |
| Gypsum | 020 | 0.2-1.5 | 11-14 |
| Pentaerythritol (PET) | 001 | 0.1-0.8 | 14-22 |

## TABLE 2

| SOLID STATE DETECTOR | CRYSTAL SPECTROMETER |
|---|---|
| **Advantages** | **Advantages** |
| 1. No x-ray focusing required | High resolving power of x-ray lines |
| 2. High solid angle possible | Quantitation-small dead time effect |
| 3. No interference from higher orders | Good peak-to-background ratio |
| **Disadvantages** | **Disadvantages** |
| 1. Needs cryogenic temperatures | Mechanical system-errors possible |
| 2. Inferior energy resolution | Only one element analysed at a time |
| 3. Non-discriminating to stray x-rays | Possible interference from higher lines |

# 6 : Surface analytical techniques

## D M POOLE

The author is with the UK Atomic Energy Authority
at Harwell Laboratory in Oxfordshire

## 1.   INTRODUCTION

For the purpose of this monograph the term 'surface'
will be taken to include material from the outermost
atomic layers to a depth of a few microns.  Techniques
to identify the elements and to study their distribution
in this material will be described roughly in order of
their depth of penetration into the sample; in addition,
some techniques which are not elemental, but which give
information on structure or bonding will be included.
Only those techniques will be considered which may
reasonably be expected to be available in the average
research laboratory, or through one of the several well-
established contract analysis units around the country.
        The principles behind the various techniques will
be described only in broad practical terms and are
illustrated by a series of schematics.  Details of the
theory can be found in the references cited at the head
of each section which are generally of a review nature.
        Wherever possible, when examples of the type of
information produced are given, they are drawn from
fields related to the theme of the meeting; two main
classes of material have been chosen for this purpose,
namely, steels and coatings.

## 2.  PRELIMINARIES

Surface analysis can be applied to a wide variety of problems arising in the use of existing materials or in the development of new products or processes. For example it may be used to identify the cause of surface staining, poor corrosion behaviour, lack of adhesion of coatings, breakdown of wear resistance etc, or may be required to determine the effectiveness of surface composition control during heat treatment, or of attempts to produce a particular composition in a protective coating being laid down by a new process.

Whatever the requirement, it is generally desirable that the physical nature of the surface in question be fully understood before any choice of analytical technique is made. Preliminary examination by optical and scanning electron microscopy should help to determine the best areas for further analysis and also help to answer various questions about the feature to be analysed: Is it a thin film perhaps giving an indication of the actual thickness through interference colours? Is the surface uniform or is there a substructure which will call for a technique with high spatial resolution? Are there breaks in the surface which will confuse the information obtained from a technique which averages over a relatively large area? Is the layer of interest thick enough to require some form of sectioning if in-depth information is needed? and so forth.

Examination in the SEM will not only answer some of the questions of the type given above, but will also give the opportunity of obtaining preliminary information about the surface and the underlying material by use of EDX analysis. Such information will be useful in deciding on the appropriate surface analysis technique and also in the interpretation of the data therefrom. However it must be remembered that the surface contamination which occurs in the course of conventional SEM examination will interfere with the more surface sensitive analytical techniques and fresh areas of sample may be needed.

Another important factor which needs to be considered is the nature of the information required: Is it purely elemental or are the valence states of interest? Is it essential to obtain quantitative figures for elemental concentrations or will purely relative values suffice?

## 3. TECHNIQUES

As indicated, these will be considered roughly in order
of their penetration into the sample.  In discussing
their relative merits various features will be taken
into account: elemental coverage, depth and lateral
resolution, detection limits etc.  The reliability of
analyses as judged by the absolute accuracy (ie the
nearness to the true value) will be loosely indicated by
describing the methods as quantitative, semi-
quantitative or qualitative - typical accuracy figures
for these categories might be within factors
respectively of 1.05, 1.25 and 2 or more.

### 3.1  SSIMS[1,2]

Static secondary ion mass spectrometry is the most
surface specific of all the techniques to be considered
here.  A low-power primary ion beam, generally of
millimetre dimensions, sputters the surface producing
ions and ion clusters which are identified by mass
spectrometry.  The rate of removal is so slow that a
monolayer on the surface can be studied for tens of
minutes or even hours; in order that adsorption of
contamination from the atmosphere should not obscure the
surface of interest, the measurement, and in general the
preparation of the layer to be studied, must be carried
out in high-quality ultra-high vacuum conditions.  For
this reason it is not likely that the technique will
find much application in the area of diagnostic
examination of practical high-temperature materials and
it will not be discussed further.  However it may be
noted that Brown and Vickerman[2] in their review of SSIMS
do in fact cite an example of its use to identify "smut"
on pickled stainless steel.

### 3.2  XPS[3,4]

X-ray photoelectron spectroscopy interrogates a surface
layer a few atom layers deep - the actual depth
depending on the details of the analysis being
performed.  An incident beam of monoenergetic soft
X-rays interacts with the electronic shells of the
target atoms and electrons are emitted whose energies

carry information about the binding energies of the
target atoms.  Since the patterns of binding energies of
all the atoms are uniquely different, measurement of
peak energies in the spectrum enables the elemental
species to be identified; elements detectable cover the
whole atomic number range from 3(Li) upwards.

Estimates of composition can be made by a
consideration of the intensities of the peaks in the
spectrum after allowing for relative elemental
sensitivity factors (Fig 1).  Detection limits are ∿ .1
to 1% and although precisions may be good, the accuracy
in the absence of suitable calibration may be such that
the method may only be considered to be semi-
quantitative.  However the relatively small range of the
sensitivity factors seen in Fig. 1 suggest that errors
will never be very large.

A particular attribute of XPS is its ability to
detect small differences in atomic binding levels
corresponding to different valence states.  This is
illustrated by Fig 2 in which the shift in the XPS peaks
when nickel is oxidised can be clearly seen.

XPS has an additional advantage over some other
techniques in its suitability for use on insulating
surfaces - for example in the examination of ceramics or
ceramic coatings.  Specimens must be compatible with the
required UHV in the instrument, however.


3.3  AES[3,4,7]

Auger electron spectroscopy has a similar information
depth to XPS as it, too, depends on examination of the
energy spectrum of relatively low-energy electrons which
have escape depths of only a few atom layers.  However
the incident radiation is in this case a beam of
electrons, focussed down to sub-micron dimensions if
required, which can create inner shell vacancies which
are then filled by electrons from outer shells; in the
Auger process, the energy released is transferred to
further electrons which may then escape from the sample
and be detected in the spectrometer of the Auger
instrument.  It will be noted that Auger emission is an
alternative process to the characteristic X-ray emission
discussed in an earlier chapter and its probability
increases with falling atomic number whilst that of
X-ray emission decreases; in this lies the reason why

Auger analysis is particularly effective for studies of
the light elements (from Be upwards) for which X-ray
emission analysis is rather difficult.  The elemental
sensitivity which can be achieved is broadly similar to
that for XPS.

As with XPS, measurement of the peaks in the energy
distribution of the emitted electrons enables the
elements present to be identified; for example Fig 3
shows the Auger spectra from two grain boundary facets
of a fractured surface of steel showing differences in
the segregation behaviour of P, Sn and Sb on the two
types of facet.  Concentrations can be estimated from
peak size using a simple formula together with tabulated
values for the relative peak intensities for the various
elements, Fig.  4; this procedure would be categorised
as semi-quantitative and the potential for errors is
emphasised by the wide range of sensitivities for which
correction is needed.  More sophisticated procedures
enable the technique to approach the quantitative.

The Auger spectra also show line shifts related to
valency of the atoms in the sample but these are
generally less prominent and precise than those in the
XPS spectrum.  Moreover the electron beam may cause
decomposition of some compounds and lead to spurious
observations.

Vacuum requirements are similar to those for XPS
and conducting specimens are generally preferred; it is
sometimes possible, however, to adjust conditions to
allow non-conductors to be analysed.


## 3.4  XPS/AES Depth Profiling

The two techniques XPS and AES, have much in common and
are frequently available together in the same UHV
instrument such as the Vacuum Generators Escalab shown
in Fig. 5.  Where information in the form of a depth
profile is required, either technique can be used in
association with an auxiliary ion gun arranged to
sputter away progressively deeper layers between
successive measurements.  The gun conditions need
careful control to produce an essentially flat-bottomed
pit so that the inherent depth resolution of the
electron spectroscopy measurement is not degraded by
slopes or irregularities.  There is, however, a strong
possibility that the true composition profile will be

upset by various ion beam effects such as preferential sputtering, ion beam mixing, ion implantation, enhanced diffusion, compound dissociation and so on. Depth profiles generated by ion sputtering must therefore always be regarded with caution in case such effects have occurred; it is particularly important to be aware that sputtering, either for depth profiling or to remove superficial contamination, may disturb the bonding characteristics which it may be desired to study using XPS.

An example of a similar method of sputter depth profiling is given in Fig. 6 where the Auger intensities of O, C, Ti and Fe have been measured over an area at the base of a series of pits of progressively greater depth in samples taken from two types of TiC coating on a steel substrate; admission of oxygen during the early stages of deposition of the carbide can be seen to have resulted in an oxygen-rich layer at the steel/carbide interface.

An alternative to building up a profile by the rather laborious processes mentioned above is to take advantage of the microprobe capability of some Auger systems. The area of interest is thinned to the necessary depth either by a mechanical means such as ball-cratering (Fig. 7), or by ion sputtering, and the Auger microprobe used to generate a depth profile by analyses at suitable positions 1, 2, 3, 4 etc, down the slope of the crater. An example of such a profile is given in Fig. 8 showing the chromated surface on a zinc coated steel; the depth scale in this example extends to tens of microns into the sample and contrasts with that in the previous example which was obtained by sputter profiling where less than half a micron had been penetrated. Profiling by sputtering is essentially rather slow ($\sim$ 5-50 nm/min) and a cratering process is generally preferable for profiling more than a micron or so.

## 3.5   SIMS[1,4,10,11]

**Dynamic** secondary ion mass spectrometry is generally implied by the acronym SIMS and, although the basic principle is the same as **static** SIMS, the rate of material removal and the lateral spatial resolution which can be achieved are both much greater. Under the more vigorous primary ion bombardment the surface is continually being sputtered away so depth profiling is inherent in the method. Vacuum requirements are less stringent than for XPS or AES, and special experimental procedures can usually be derived to deal with insulating samples.

The best depth resolution which can normally be achieved is $\sim$ 10 nm, but to demonstrate this when an area of perhaps 100 µm dimensions is being analysed, requires a very high degree of perfection in the sample, such as can normally only be seen in semiconductor materials; an example is shown in Fig. 9 where alternating layers of GaAs and GaAlAs are being profiled. Much less sharp transitions from one region to another will be seen when profiling through, say, an oxide layer on steel as a result of the inherent lack of perfection in the sample, and place to place sputtering variations which will arise due to grain orientations, impurities etc; this can be seen in Fig. 10 which shows a conventional profile through an oxide obtained using a primary argon ion beam, and also illustrates the use of gallium ions as the primary beam with sub-micron potential resolution.

SIMS is not only used to generate depth profiles from relatively large areas tens of microns in size - which it does at a relatively rapid rate of up to a micron in a few minutes - but it can also be used to give laterally resolved information. This can be achieved in either of two ways: (a) in an imaging instrument such as the Cameca 3F or 4F, which is essentially an ion microscope where the ions emitted from the field of view are mass selected and individually formed into an image with lateral resolution of $\sim$ 1 µm, or (b) in a scanning instrument such as the VG Ionex SIMSLAB, where a finely focussed primary beam of gallium ions for example, is scanned over the field and selected ion images formed in a manner exactly analogous to the electron image in a SEM. The imaging mode can be very useful for checking that there are no gross non-uniformities in elemental distribution (eg second phases, or impurity particulates) which can give rise to distortions in the depth profiles, or can give direct indication of depth distributions by viewing the edge of a crater or a conventional cross section through thicker features. Fig. 11 shows an ion image of the cerium distribution through a NiO scale which was obtained during a study of

241

the way in which certain reactive elements such as cerium can be used to reduce the high temperature oxidation of nickel alloys or steels; a corresponding depth profile is shown in Fig. 12.

The last figure gives an indication of one of the main drawbacks of SIMS, namely that it is essentially only a qualitative technique and generally results are reported as "intensity-arbitrary units" rather than concentration. The main difficulty arises from the "matrix effects" which occur resulting in sometimes very large departures from a simple "signal proportional to concentration" relationship; for example the yield of $Ni^+$, $Cr^+$ and $Fe^+$ ions from an oxide can exceed those from the metal substrate by as much as an order of magnitude. A further complication lies in the huge variations in relative ion yields that are seen across the periodic table; Fig. 13 shows this variation in the case of positive ion yields from elements bombarded by oxygen or caesium, and differences of several orders of magnitude can be seen. Thus there are two related reasons why interpretation of SIMS data can be difficult; theoretical models can be applied with some success in an attempt to overcome the problems, but the most reliable approach is calibration using suitable standards or by applying some other quantitative method.

Apart from profiling speed and lateral resolution noted earlier, SIMS has other particular advantages which give it an important place in the spectrum of available techniques. Inherent in a mass spectrometric method is the ability to distinguish different isotopes and this can sometimes be valuable in naturally occurring circumstances, but more especially in experiments involving tracers. An excellent example of the latter is the use of $^{18}O_2$ to label an oxidising environment at a particular time in a study of a high temperature oxidation; subsequent profiling of the $^{18}O$ and $^{16}O$ species in the oxide will give valuable information on the mechanism of oxygen transport in the growing scales.

Other advantageous features are the phenomenal sensitivity for many elements. Fig. 13 showed elements such as Na, Al etc giving very high positive ion yields under $O^-$ ion bombardment, and similarly high yields of negative ions from many elements such as sulphur and gold can be obtained with a $Cs^+$ primary beam. These high yielding elements can be detected at or below the ppm level with parts per billion being achieved in certain cases.

## 3.6  GDOS and GDMS[16,17]

Surface analysis versions of glow discharge optical spectroscopy and glow discharge mass spectrometry have in common the creation of an argon ion plasma with the sample itself forming the cathode and being eroded away to give an inherently depth profiling technique.  The sample is usually required to be flat and large enough ($\sim$ 20 mm) to be sealed against an aperture in the discharge source body; the area which is actually profiled is $\sim$ 10 mm dia and there is no lateral resolution capability.  Erosion rates from .01 $\mu$m/min to 10 $\mu$m/min are reported and these offer a range from precise profiling with good depth resolution $\sim$ 1 nm, to extremely rapid and deep profiling with depth resolution deteriorating to perhaps $^1$/10's of $\mu$m as the depth increases.

The principles of analysis by optical spectroscopy have been considered in an earlier paper and it is sufficient to emphasise here that the technique is highly sensitive (ranging from .1 to .001$^W$/o), quantitative with appropriate calibration, capable of simultaneous analysis of tens of elements, and potentially very rapid.  The speed and versatility of the method is illustrated in Fig. 14 where detailed six element depth profiles through oxide layers several microns thick on Hastelloy have been obtained in 5 mins.

The mass spectrometric variant clearly has some of the characteristics of the SIMS technique already discussed:  specifically high sensitivity and isotopic discrimination.  However the secondary ion production by plasma discharge very largely overcomes the enormous yield variations which so plague SIMS; this is illustrated in Fig. 15 showing a variety of elemental yields from a range of materials with a total spread of only one order of magnitude – compare this to the five orders seen in the case of SIMS.  The relative freedom from matrix effects also means a more direct relationship between intensity and concentration when passing from an oxide to a metallic substrate for

example. Analysis without standards should be at least
semi-quantitative and this, coupled with the rapid
profiling capability make GDMS a very powerful
technique.


## 3.7 EMPA[10,20,21]

Electron microprobe analysis used in a conventional
manner has been covered in earlier papers; the normally
quoted depth resolution of 1 μm is widely accepted and
does characterise normal use of the instrument. However
the electron microprobe does have a place in surface
analysis since it is a truly quantitative instrument and
it can be operated in such a way that the depth analysed
can be reduced to 100 nm or less. These characteristics
may be used to analyse thin films which pose particular
problems to other surface techniques; for example the
features to be examined may be too small for XPS, and
may be insulating so that AES may not be used. A
special case of this low penetration use of the
microprobe is the calibration of high resolution, but
purely qualitative SIMS depth profiles; Fig. 16
illustrates this in the case of determination of
phosphorous in a thin $SiO_2$ film.
   When surface films are thick enough, and
inhomogeneous enough to call for high spatial resolution
examination of cross-sections, then the electron
microprobe is frequently the preferred technique; with
modern control systems multi-element high resolution,
semi-quantitative distribution maps can be generated in
reasonable time periods ∿ hours - if necessary in
overnight runs. For fully quantitative information
point by point line scans can be generated and the X-ray
intensity plots, such as those in Fig. 17, can readily
be converted to concentration profiles.


## 3.8 LIMA or LAMMA[4,22]

Laser ionisation mass analysis or laser microprobe mass
analysis, the alternative trade name, is another probe
technique for the analysis of solids with lateral
resolution of the order of 2 μm and a possible depth
resolution ∿ 100 nm. However it differs greatly from
EMPA in that it is destructive as each laser shot

creates a melted crater and inevitably causes some
disturbance of the composition of the surrounding
material. The ions produced in the vapour cloud are
collected in a time-of-flight mass spectrometer (ie
essentially simultaneously), the peaks identified, and
their areas estimated for comparison purposes. The
technique tends to suffer from poor reproducibility but
may perhaps be described as semi-quantitative through
the use of calibration standards. The elemental
coverage is total, ranging from H, mass 1, upwards and
sensitivites are down to parts per billion dependent on
element, laser power etc.

In common with those from SIMS, the mass spectra
contain molecular ion fragments which may be seen merely
as a complication if the objective is purely elemental
identification and estimation; the molecular ion
species, however, carry information about the molecular
features of the target, confused to an extent by
molecular ions actually formed in the vaporisation
process, and thus may be able to give valuable chemical,
as distinct from elemental, information.

At low laser powers the analysis becomes more
surface sensitive and can be used to study surface
chemisorption effects. Clarke et al[22] have compared the
intensity of the $H_2O^+$ and other ions originating from
the chemisorbed water on a variety of different oxides
of iron - a typical spectrum is given in Fig. 18.


## 3.9  RBS and ERDA[10,17,23]

Rutherford backscattering spectrometry and elastic
recoil detection analysis share a feature which
distinguishes them from all the techniques discussed so
far: the primary beam interacts with the sample to a
depth of 1 or 2 μm and the continuous signal which is
produced can be deconvoluted to yield information about
the atoms present and their distribution in depth - ie
we have simultaneous multi-element depth profiling
without the removal of successive layers of the target.

In the case of RBS $He^+$ ions bombard the target and
their backscattered energy spectrum carries information
about the mass and position of the target atoms which
have scattered the beam (Fig. 19). The method is best
for heavy atoms in a light matrix and can have a depth
resolution ∿ 20 nm. Detection sensitivites are high

(down to ppm levels) and the technique is at least semi-quantitative.

The situation is rather the reverse with ERDA in which a heavy ion beam scatters atoms out of the target and their energy distribution (dependent on the atom mass and its origination depth) is used to identify the atoms and deduce their depth distributions. The technique covers the very light elements from H to F with a sensitivity typically $\sim$ .1$^a$/o and a depth resolution similar to RBS; the method is also at least semi-quantitative. Although sometimes described as a non-destructive technique, the ion beam in fact causes considerable damage and is implanted in significant quantities into the specimen.

The advantages of simultaneous, nominally non-destructive, multi-element depth profiling for elements down to H, are obvious. However there may be problems with elemental interferences which prevent the deconvolution of the spectra and the methods may not be applicable in all cases. Nor can they be found in all laboratories as the high-energy incident beams (several MeV) require the availability of powerful accelerator installations. Lack of spatial resolution is a drawback which is steadily being limited with progress towards microbeams of 1 μm dimensions; the practical limit at present appears to be $\sim$ 3 μm.

3.10  NRA[4,25]

Nuclear reaction analysis is also an accelerator-based technique, but its potential for built-in depth profiling is poor when compared to RBS or ERDA. MeV beams of light ions undergo resonant interactions with specific nuclei in the sample and other particles are emitted which can be used to identify and quantify the elements in the target. The technique is particularly powerful for the detection of the light elements such as carbon (through a $^{12}$C(d,p) reaction for example) and is probably one of the best for high sensitivity, quantitative measurements. Not being surface specific it is mainly of value for the examination of sections where a resolution of several microns (dictated by the minimum achievable beam size) is acceptable. Such an application is shown in Fig. 20 where the carbon distribution in iron has been measured to a distance below the surface of 100 μm.

246

## 3.11  LRS[10,26]

Laser raman spectroscopy is not an elemental analysis
technique as the preceding techniques have been, but is
one which can detect vibrational frequencies of
molecular bonds in the sample.  A high frequency light
source in the form of a laser beam is focussed onto the
sample and the scattered spectrum contains peaks shifted
from the incident frequency by an amount equal to the
molecular vibration frequency.  The positions of these
peaks provide  a means of identifying particular
molecular bonds and hence structures or compounds; the
frequency characteristics of the bonds in possible
sample components must be known from existing
compilations or from measurement on appropriate
standards.  Lateral resolution can be $\sim$ 1 μm for the
microprobe form of the technique, but depth of analysis
depends on the optical characteristics of the sample.

An example of the identification of oxide species
on corroded steel is shown in Fig. 21 where the same
cross-section previously examined by EMPA (Fig. 17), has
been probed at various positions and the spectra
recorded; the various oxide peaks are identified in the
figure and these relate directly with the elemental
information determined earlier.

As both incident and emergent beams are light, the
technique has a potential for in-situ examination of
surfaces.  The light can be directed through windows or
along optical fibres to the areas to be sampled, without
the need to prepare a specimen, or insert it into a
vacuum chamber as with most of the preceding
techniques.

## 3.12  XRD[10,28]

X-ray diffraction is not generally considered to be a
surface technique.  It is possible however by suitable
choice of primary beam (eg Cr-Kα instead of the more
penetrating Cu-Kα), and by operating at glancing angles,
to obtain structural information from quite thin surface
layers.  An example of this approach is given in Fig. 22
showing the development of various oxide phases during
the oxidation of steel in $CO_2$ at various temperatures;
the layers were well below 1 μm thickness at the lower
temperatures but were nevertheless quite readily
identified.

## 4. SUMMARY

It should be apparent from the foregoing that in many instances when surface analysis is required there will be more than one technique which can be used - selection may well then be made on grounds of availability and cost rather than fine differences in technical merit. What has not so far been emphasised is the frequent need to use more than one technique in order to fully characterise a situation; an example of this was given in the case of combining sensitivity and profile precision of SIMS with the quantification capability of EMPA.

## ACKNOWLEDGEMENTS

Thanks are due to Mrs Pat Hetherington for her able assistance in the preparation of the text, and to various colleagues at the Harwell Laboratories for permission to use illustrations from their work.

1. A. BENNINGHOVEN, F.G. RÜDENAUER and H.W. WERNER: "Secondary Ion Mass Spectrometry: Basic Concepts, Instrumental Aspects, Applications and Trends"; 1987, New York, John Wiley & Sons.

2. A. BROWN and J.C. VICKERMAN: Surf. Interface Anal., 1984, **6**, 1-14.

3. D. BRIGGS and M.P. SEAH (eds.): "Practical Surface Analysis by Auger and X-ray Photoelectron Spectroscopy"; 1983, Chichester, John Wiley & Sons.

4. H.W. WERNER and R.P.H. GARTEN: Rep. Prog. Phys., 1984, **47**, 221-344.

5. C.D. WAGNER: Anal. Chem., 1972, **44**, 1050-1053.

6. P.R. CHALKER: Private Communication.

7. H.E. BISHOP, Chap. 4: "Methods of Surface Analysis"; 1988, Cambridge University Press.

8.   L.E. DAVIS, N.C. MacDONALD, P.W. PALMBERG, G.E.
     RIACH and R.E. WEBER: "Handbook of Auger Electron
     Spectroscopy", 2nd Edn.; 1976, Minnesota, Physical
     Electronics Div., Perkin-Elmer Corp.

9.   A. PAN and J.E. GREENE: Thin Solid Films, 1982,
     **97**, 79-89.

10.  Metals Handbook, Ninth Edition, Vol. 10, "Materials
     Characterization", 1986, American Society for
     Metals, Ohio.

11.  H.W. WERNER: "Quantitative Secondary Ion Mass
     Spectrometry A Review"; Surf. Interface Anal.,
     1980, **2**, 56-74.

12.  H.E. BISHOP: Private Communication.

13.  H.E. BISHOP, D.P. MOON, P. MARRIOTT and
     P.R. CHALKER: Harwell Laboratory report
     AERE-R13081, 1988, Submitted to Surf. Interface
     Anal.

14.  D.P. MOON and H.E. BISHOP: SIMS VI, 1987,
     Versailles, to be published.

15.  H.A. STORMS, K.F. BROWN and J.D. STEIN: Anal.
     Chem., 1977, **49**, 2023 as reproduced in Atomica
     leaflet.

16.  W.W. HARRISON, K.R. HESS, R.K. MARCUS, F.L. KING:
     Analytical Chem., 1986, **58**, 341A-356A.

17.  C.W. MAGEE and L.R. HEWITT: RCA Review, 1986,
     **47**, 162-185.

18.  H. NICKEL: Proceedings of the 13th Colloquium on
     Metallurgical Analysis, Vienna, May 1987,
     Mikrochim. Acta [Wien], 1987, **1**, 5-47.

19.  N.E. SANDERSON, E. HALL, J. CLARK, P. CHARALAMBOUS
     and D. HALL, Ref. 18, p. 275.

20.  K.F.J. HEINRICH: "Electron Beam X-ray
     Microanalysis"; 1981, New York, Van Nostrand
     Reinhold.

21. V.D. SCOTT and G. LOVE: "Quantitative
    Electron-Probe Microanalysis"; 1983, Horwood
    (Ellis).

22. N.S. CLARKE, J.C. RUCKMAN and A.R. DAVEY: Surf.
    Interface Anal., 1986, **9**, 31-40.

23. W.K. CHU, J.W. MAYER and M.A. NICOLET:
    "Backscattering Spectrometry; 1978, Academic
    Press.

24. P.M. READ, Private Communication.

25. G. AMSEL, J.P. NADAI, E. D'ARTEMARE, D. DAVID,
    E. GIRARD and J. MOULIN: Nucl. Instrum. Methods,
    1971, **92**, 481.

26. D.J. GARDINER, M. BOWDEN and P.R. GRAVES: Phil.
    Trans. Roy. Soc. Lond. **A**, 1986, **320**,
    295-306.

27. B.A. BELLAMY: European Spectroscopy News, 1987,
    **74**, 12-19.

28. B.D. CULLITY: "The Elements of X-ray Diffraction",
    2nd Edn.; 1978, Reading, Massachusetts,
    Addison-Wesley.

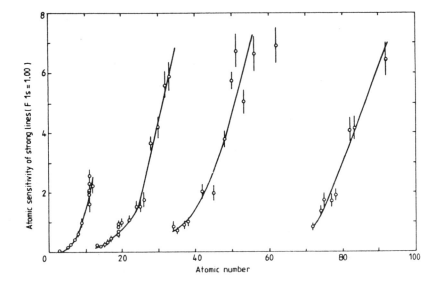

1.  XPS atomic sensitivity factors of the elements
    relative to the F 1s line (from Wagner[5]).

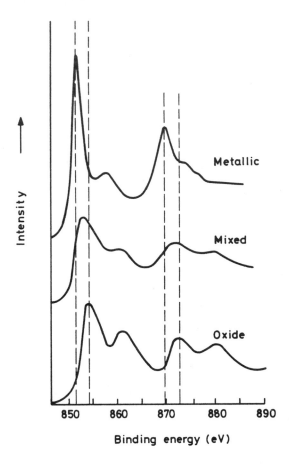

2.  XPS spectra of nickel showing the binding energy
    differences between the metallic and oxidised
    state[6].

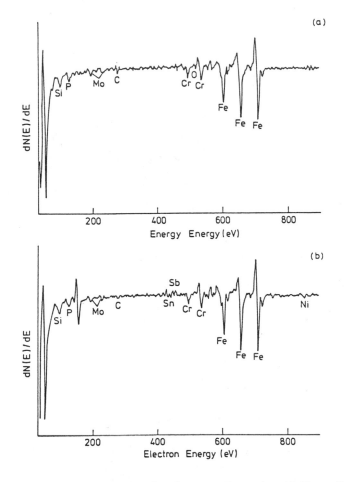

3.   AES spectra obtained from Type A and Type B
     intergranular facets of fractured steel[7].

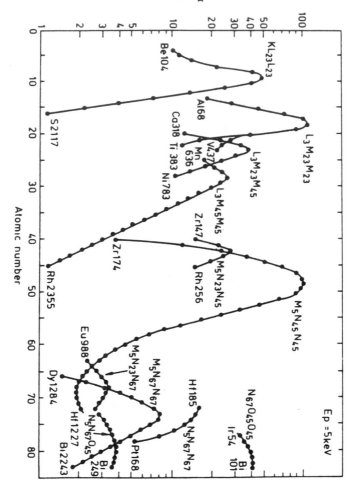

4. Relative sensitivity factors for derivative spectra in AES compiled from Davis et al[8]. The numbers at the ends of the curves indicate the energies of the peaks and the elements at these points[3].

5.    Vacuum Generators ESCALAB at Harwell.

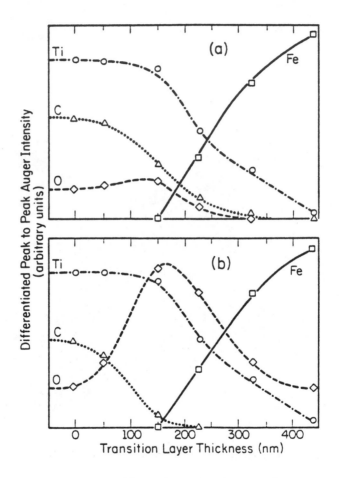

6.  AES depth profiles through TiC coatings on steel
    formed under (a) low and (b) high oxygen partial
    pressure[9].

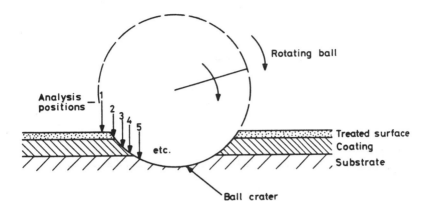

7.   Sectioning by ball cratering followed by analysis
     at various depths 1, 2, 3 etc.

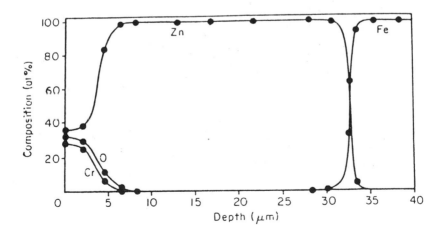

8.   Depth profile through ball cratered zinc coating on
     steel[7].

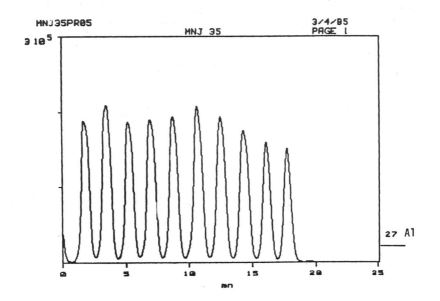

9.  SIMS depth profile through alternating 10 nm layers of GaAs and GaAlAs. $^{27}Al^{+}$ signal shows good resolution to 200 nm[12].

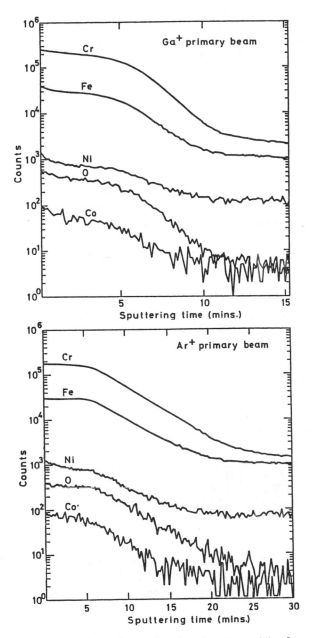

10. Depth profiles through the inner oxide layer on a
stainless steel specimen using the gallium ion beam
and an argon ion beam in the IMS 3F[13].

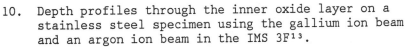

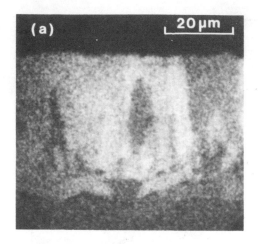

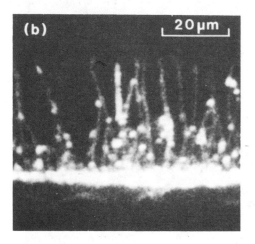

11.  (a) $Ni^+$(58) and (b) $CeO^+$(156) secondary ion images
     of a ceria-dispersed NiO scale in transverse
     section.  (Cs primary beam)[14].

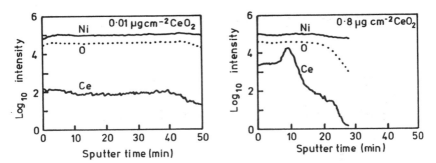

12. Depth profiles of ceria particle-dispersed nickel
    oxide scales formed after 142h at 900°C.
    (Cs primary beam)[14].

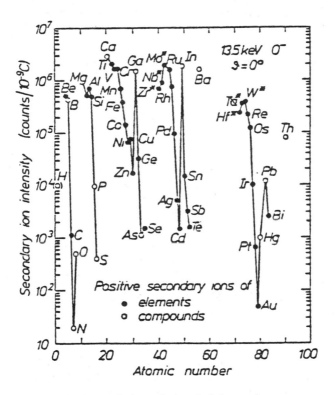

13. Relative positive ion yields under oxygen or
    caesium ion beams at normal incidence[15].

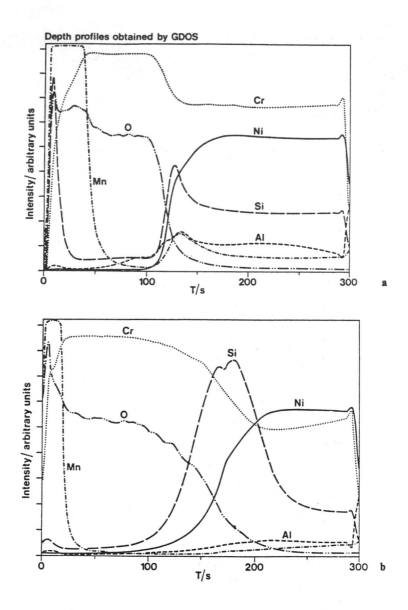

Depth profiles obtained by GDOS

14. Depth profiles of two oxidised HASTELLOY X
specimens, determined using GDOS after oxidation
experiments at different oxygen partial
pressures[18].

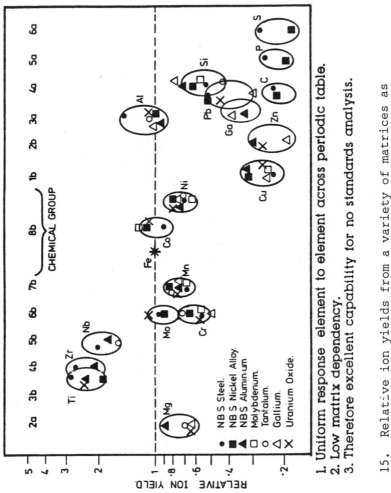

1. Uniform response element to element across periodic table.
2. Low matrix dependency.
3. Therefore excellent capability for no standards analysis.

15. Relative ion yields from a variety of matrices as measured in the VG 9000 GDMS instrument[19].

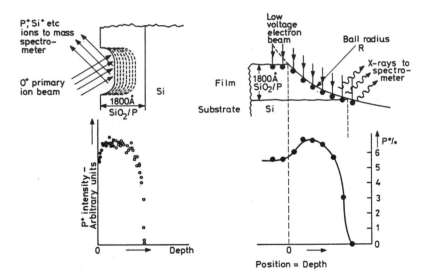

16. Illustrating the use of EMPA on ball cratered sample to calibrate a high sensitivity depth profile obtained by SIMS.

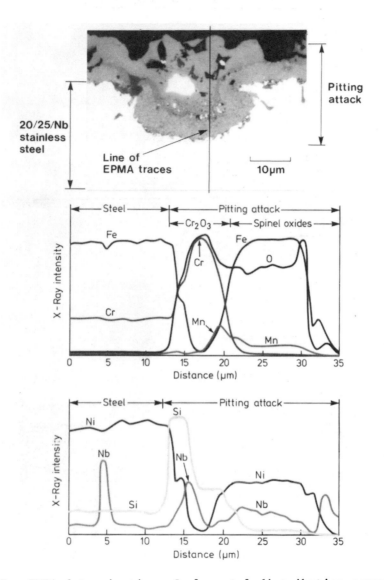

17. EMPA determination of elemental distribution across
    the pitting attack of a 20/25/Nb stainless steel at
    900°C along the traverse shown in the optical
    micrograph.

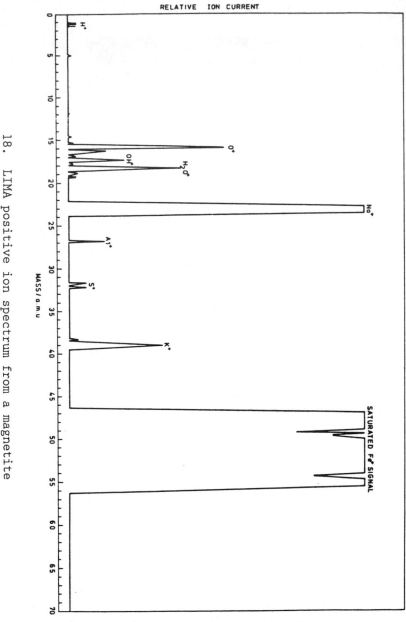

18. LIMA positive ion spectrum from a magnetite sample[22].

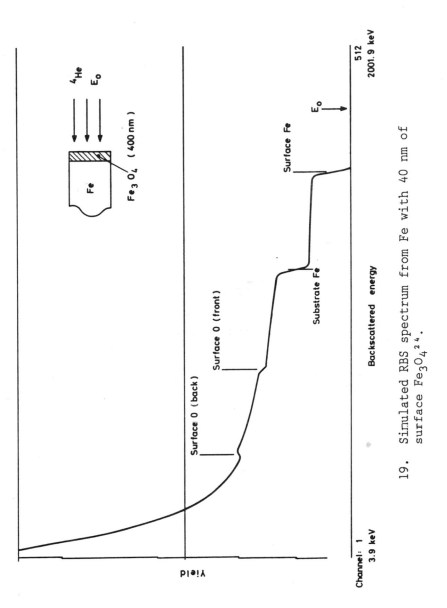

19. Simulated RBS spectrum from Fe with 40 nm of surface $Fe_3O_4{}^{24}$.

267

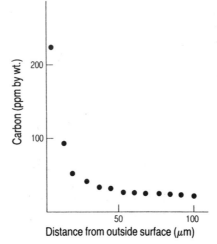

20. Determination of carbon depth profile in section through an iron sample using the $^{12}C$ (d,p) reaction.

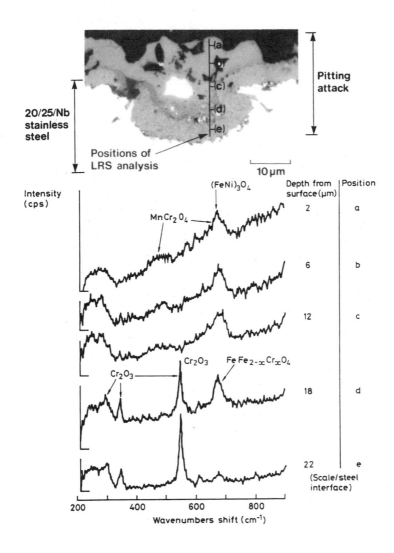

21. Raman spectra obtained at the positions shown in the optical micrograph across the pitting attack of 20/25/Nb stainless steel at 900°C[27].

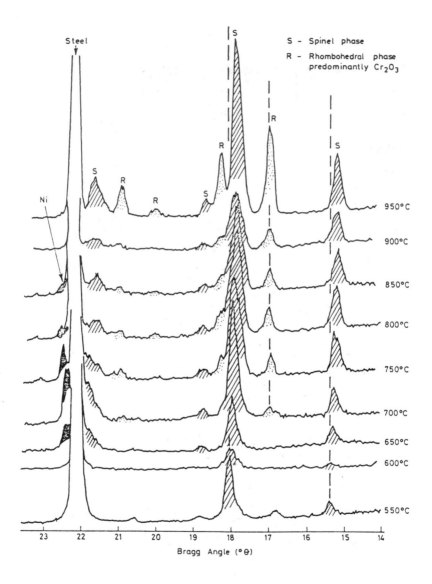

22. X-ray diffraction patterns of the oxide films
    formed on 20/25/Nb stainless steel during 25h
    exposure to $CO_2$ + 4% CO + 350 vpm $CH_4$ + 300 vpm
    $H_2O$ + 400 vpm $H_2$ at 550-950°C[29].